中国石油和化学工业优秀出版物奖（教材）

江苏省高等学校精品教材

化工专业英语

第二版

English for Chemical Engineering and Technology

张小军　主编
周志刚　主审

化学工业出版社
·北京·

内 容 简 介

本书是为了提高化工技术类专业学生化工英语信息获取能力和应用能力，培养与职业能力结构要求相一致的高素质技术技能型人才而编写的。全书选文均来自原版化工英文文献，侧重化工实际工艺及技术操作，覆盖面广、难度适中。内容包括化工专业英语基础（Unit One Basis of English for Chemical Engineering and Technology）、石油化学工业（Unit Two Petrochemical Industry）、聚合物化学（Unit Three Polymer Chemistry）、安全工程（Unit Four Safety Engineering）、化学工程技术（Unit Five Chemical Engineering and Technology）、精细化学品（Unit Six Fine Chemicals）、分析化学（Analytical Chemistry）、生物化工（Unit Eight Biochemical Engineering）、环境污染及治理（Unit Nine Environmental Pollution and Control）。其中"化工专业英语基础"部分，对提高读者化工专业英语的阅读、会话和写作能力有很高的实用价值。

本书可作为化工类专业高等职业院校学生的教材，也可作为化工领域广大科技人员的参考书。

图书在版编目（CIP）数据

化工专业英语/张小军主编. —2 版. —北京：化学工业出版社，2023.2 （2024.8重印）
高等职业教育教材
ISBN 978-7-122-42650-5

Ⅰ.①化⋯ Ⅱ.①张⋯ Ⅲ.①化学工程-英语-高等学校-教材 Ⅳ.①TQ02

中国国家版本馆 CIP 数据核字（2023）第 001008 号

责任编辑：窦　臻　林　媛　　　　　装帧设计：史利平
责任校对：田睿涵

出版发行：化学工业出版社（北京市东城区青年湖南街 13 号　邮政编码 100011）
印　　装：河北延风印务有限公司
787mm×1092mm　1/16　印张 13½　字数 330 千字　2024 年 8 月北京第 2 版第 2 次印刷

购书咨询：010-64518888　　　　　　　售后服务：010-64518899
网　　址：http://www.cip.com.cn

凡购买本书，如有缺损质量问题，本社销售中心负责调换。

定　　价：39.00 元　　　　　　　　　　　　　　　　版权所有　违者必究

PREFACE 前言

科学技术本身的性质要求科技专业英语（english for special science and technology）与专业内容相互配合、相互一致。国际上诸多化工专业资料都是以英文形式写成的，化工技术领域的国际交流主要是通过英语进行的，化工专业英语（english for chemical engineering and technology）既是传播信息的媒介，又是通向化工世界的桥梁和不可缺少的有力工具。因此，化工专业英语的学习与运用，对于我国化工专业技术人员向国外介绍国内化工技术、吸收消化国外先进的化工技术都是十分必要的。

南京科技职业学院以普通英语、专业英语和专业双语教学的"大英语"教学观为指导，结合学校近30年化工专业英语教学实践，对教学内容进行了系列整合与更新，巩固学生的化工专业能力、提升英语应用能力、拓展职业生涯的可持续发展能力，使之能够适应现代化工科技的发展趋势，满足21世纪我国对高层次化工技术人才的需求。本教材以编者团队编写的"江苏省高等学校精品教材"（2011年）《化工专业英语》为基础，全书选文侧重化工实际工艺及技术操作，内容涵盖了石油化学工业、聚合物化学、安全工程、化学工程技术、精细化学品、生物化工、环境污染及治理等，同时，根据化工类专业的特点，分别以独立的单元详细介绍了化学工程、有机化工、工业催化、高分子化工等专业相关的基础理论、生产技术和最新发展，内容均选自原版化工英文文献材料。为提高学生化工专业英语的阅读、翻译和写作能力，在本书中还专门介绍了专业英语的基础知识、特点、学习方法等，期望使用本书的读者通过"化工专业英语"课程的学习，基本具备外资企业化工专业技术人员所需达到的英语应用能力，能够借助英语开展化工专业领域的交流、应用文起草、专业文献阅读与翻译工作，较好地服务化工专业实践工作，拓展职业生涯。本教材第一版被评为中国石油和化学工业优秀出版物奖（教材奖）一等奖（2016年）。

随着"三教"改革的深入，"化工专业英语"课程建设如何置于新时代职业教育改革发展的整体背景下进行，这是首要考虑的问题。《化工专业英语》此次修订在保留了前版特色的同时，精简了内容篇幅，新增了以微课为主的数字资源，以二维码形式链接在教材中。新增了两个特色栏目"History, Inheritance and Development""Practice and Training"，以期提高专业英语课程思政实效，从行业历史、传承和发展中找到中国精神、工匠精神、劳模精神，落实了党的二十大报告提出的"培养造就大批德才兼备的高素质人才"的要求，增加了化工实验、实训内容篇幅，增强学生英文环境下的专业实践实操能力。

本教材主编张小军教授，先后从事过大学英语、化工专业英语及化工专业课三类课程的教学工作，负责本教材的编写逻辑与结构，并且编写了第 1 单元——化工专业英语基础［Unit One Basis of English for Chemical Engineering and Technology（ECET）］。第 2 至 9 单元的编写者分别是南京科技职业学院的王一男（Unit Two Petrochemical Industry 石油化学工业）、薛华玉（Unit Three Polymer Chemistry 聚合物化学）、刘健（Unit Four Safety Engineering 安全工程、Unit Five Chemical Engineering and Technology 化学工程技术）、胡瑾（Unit Six Fine Chemicals 精细化学品）、徐琳（Unit Seven Analytical chemistry 分析化学）、沈建华（Unit Eight Biochemical Engineering 生物化工）和王瑞（Unit Nine Environmental Pollution and Control 环境污染及治理）。王一男具体负责校稿、排版工作。本教材由天津大学博士生导师周志刚教授担任主审。此外，在本书修订过程中得到了化工职教集团兄弟院校、化工行业企业的工程技术人员以及化学工业出版社的大力支持，在此一并表示衷心感谢！

由于编者水平有限，疏漏和不妥之处在所难免，恳请读者提出宝贵意见。

编者

二〇二三年二月

第一版前言

科学技术本身的性质要求专业英语（english for special science and technology）与专业内容相互配合、相互一致。国际上70%以上的化工专业资料都是以英文形式写成的，化工技术领域的交流主要是通过英语进行的，化工专业英语（english for chemical engineering and technology）既是传播信息的媒介，又是通向化工世界的桥梁和不可缺少的有力工具。因此，化工专业英语的学习与运用，对于我国化工专业技术人员介绍国内化工技术、吸收消化国外先进的化工技术都是十分必要的。

南京科技职业学院（原南京化工职业技术学院）以普通英语、专业英语和专业双语教学的"大英语"教学观为指导，结合学校20多年化工专业英语教学实践，对教学内容进行了系列整合与更新，巩固学生的化工专业能力、提升英语应用能力、拓展职业生涯的可持续发展能力，使之能够适应现代化工科技的发展趋势，满足21世纪我国对高层次化工技术人才的需求。本教材以编者团队编写的"江苏省高等学校精品教材"（2011年）——《化工专业英语》为基础，全书选文侧重化工实际工艺及技术操作，内容涵盖了石油化学工业、聚合物化学、安全工程、单元操作、化学工程技术、精细化学品、生物化工、环境污染及治理等，同时，根据化工类专业的特点，分别以独立的单元详细介绍了化学工程、有机化工、工业催化、高分子化工等专业相关的基础理论、生产技术和最新发展，内容均选自原版化工英文文献材料。为提高学生化工专业英语的阅读、翻译和写作能力，在本书中还专门介绍了专业英语的基础知识、特点、学习方法以及专业英语翻译及写作等，期望使用本书的读者通过"化工专业英语"课程的学习，具备外资企业化工专业技术人员所需达到的英语应用能力，能够借助英语开展化工专业领域的交流、工作过程应用文起草、专业文献阅读与翻译工作，较好地服务化工专业实践工作，拓展职业生涯。

本教材主编张小军教授，先后从事过大学英语、化工专业英语及化工专业课三类课程的教学工作，负责本教材的编写逻辑与结构设置，并且编写了第1单元——化工专业英语基础（Unit One Basis of English for Chemical Engineering and Technology）的第一、二、四、五课内容及化工专业英语词汇。第2至第10单元的编写分工如下：南京科技职业学院的王一男老师编写了 Unit Two Petrochemical Industry 石油化学工业、薛华玉老师编写了 Unit Three Polymer Chemistry 聚合物化学、刘健老师编写了 Unit Four Safety Engineering 安全工程、吴莉莉老师编写了 Unit Five Unit Operation 单元操作、于清跃老师编写了 Unit Six Chemical Engineering and Technology 化学工程技术、胡瑾老师编写了 Unit Seven Fine Chemicals 精细化学品、徐琳老师编写了 Unit Eight

Introduction of Quantitative Analysis 分析化学、沈建华老师编写了 Unit Nine Biochemical Engineering 生物化工、王瑞老师编写了 Unit Ten Environmental Pollution and Control 环境污染及治理、文冰老师编写了第 1 单元——化工专业英语基础（Unit One Basis of English for Chemical Engineering and Technology）的第三课内容及化工专业英语拓展应用的Ⅱ、Ⅲ、Ⅳ、Ⅴ部分。王一男老师具体负责校稿、排版工作。本教材由天津大学博士生导师周志刚教授担任主审。此外，在本书编写过程中得到了南京科技职业学院化学工程系与应用化学系、兄弟院校、化工行业企业的工程技术人员以及化学工业出版社的大力支持，在此一并表示衷心感谢！

由于编者水平有限，疏漏和不妥之处在所难免，恳请提出宝贵意见。

编者
二〇一五年五月

CONTENTS 目录

Unit One
Basis of English for Chemical Engineering and Technology (ECET)
化工专业英语基础

Lesson One　Outline of Learning ECET
化工专业英语学习概要　　1

Lesson Two　Characteristics of ECET
化工专业英语特点　　2

Lesson Three　Vocabulary of ECET
化工专业英语词汇　　3

Lesson Four　Writing and Speaking Frequently Occurred in ECET
化工专业英语常用写作与会话　　6

Lesson Five　Grammar Frequently Used in ECET
化工专业英语常用语法　　13

Unit Two
Petrochemical Industry
石油化学工业

Lesson One　Petroleum Refinery Distillation
石油炼制　　23

Lesson Two　Catalytic Cracking
催化裂化　　27

Lesson Three　Thermal Cracking of Hydrocarbons
热裂解制乙烯　　33

History, Inheritance and Development　　40

Practice and Training　　42

Unit Three
Polymer Chemistry
聚合物化学

Lesson One　Basic Concepts of Polymers
聚合物的基本概念　　45
Lesson Two　Polymer Structure and Physical Properties
聚合物的结构及其物理性能　　48
Lesson Three　Applications of Polymers
聚合物的应用　　52
History, Inheritance and Development　　56
Practice and Training　　57

Unit Four
Safety Engineering
安全工程

Lesson One　A Brief History of Safety
安全工程简史　　58
Lesson Two　The System Safety Process
系统安全过程　　62
Lesson Three　Basic Principles for Controlling Chemical Hazards
管理化学危险的基本原则　　69
History, Inheritance and Development　　77
Practice and Training　　78

Unit Five
Chemical Engineering and Technology
化学工程技术

Lesson One　Primary Principles of Unit Operation
单元操作基本原理　　80
Lesson Two　Fluid Flow Phenomena
流体流动现象　　86
Lesson Three　Chemical Engineering
化学工程　　90

Lesson Four   The Anatomy of a Chemical Manufacturing Process
化工生产过程分解   95
Lesson Five   Catalysts for Industrial Processes
工业催化剂   99
History，Inheritance and Development   103
Practice and Training   103

Unit Six
Fine Chemicals
精细化学品

Lesson One   A Brief Introduction of Fine Chemicals
精细化学品简介   105
Lesson Two   Coating
涂料   110
Lesson Three   Classification and Application of Surfactants
表面活性剂的分类与应用   115
Lesson Four   Plasticizers
增塑剂   120
History，Inheritance and Development   125
Practice and Training   126

Unit Seven
Analytical Chemistry
分析化学

Lesson One   How to Use Analytical Apparatus
如何使用分析仪器   127
Lesson Two   Titrimetric Methods
滴定分析   133
Lesson Three   Gas Chromatography
气相色谱   138
History，Inheritance and Development   143
Practice and Training   144

Unit Eight
Biochemical Engineering
生物化工

Lesson One　Introduction to Biochemical Engineering
生物化工简介　　145
Lesson Two　Molecular Structure of Nucleic Acid—A Structure for Deoxyribose Nucleic Acid
核酸的分子结构——脱氧核糖核酸的结构　　150
Lesson Three　Biochemical Reaction
生化反应　　155
History，Inheritance and Development　　159
Practice and Training　　160

Unit Nine
Environmental Pollution and Control
环境污染及治理

Lesson One　Water Pollution and Pollutants
水污染及污染物质　　161
Lesson Two　Air Pollution and Major Air Pollutants
大气污染及主要的大气污染物　　170
Lesson Three　Sources and Types of Solid Wastes
固体废物的来源和类别　　179
History, Inheritance and Development　　184
Practice and Training　　185

Appendix
Glossary
词汇表

Reference
参考文献

Unit One

Basis of English for Chemical Engineering and Technology (ECET)

化工专业英语基础

▶▶▶▶

Lesson One Outline of Learning ECET
化工专业英语学习概要

Part One Concept of ECET
化工专业英语概念

科学技术本身的性质要求专业英语（english for special science and technology）与专业内容相互配合、相互一致，这就决定了专业英语与普通英语（common english or general english or ordinary english）具有很大差异。专业英语的主要特点是它具有很强的专业性，懂专业的人用起来得心应手，不懂专业的人用起来则困难重重。

就化工专业而言，由于历史的缘故，目前，国际上化工技术的交流主要是使用英语，70%以上的化工专业资料都是以英文形式写成的。因此，化工专业英语的学习与运用，对于化工专业技术人员介绍国内化工技术，吸收、消化国外先进的化工技术，参与化工技术交流是十分重要的。今天，我们一起开启"化工专业英语"的学习航程也是完全有必要的。"化工专业英语"（english for chemical engineering and technology）是一门化工类专业必修课，其前修课程一般是公共英语和化工类专业课程，后修课程为化工专业双语课程和化工专业方向课程。该课程是以化工类专业内容和化工企业专业实践的视角，在公共英语、专业英语与双语教学课程这个连续化的"大英语"教学体系中，定位"化工专业英语"课程内容及其教学工作，它以化工工作过程为导向，以英语的专业应用能力为主线，以化工专业内容为载体，进行模块化课程教学改革，培养学生良好的化工专业领域工作过程的英语应用能力，服务于化工专业工作实践，拓展职业生涯。

Part Two Learning Targets of ECET
化工专业英语的学习目标

1. 总体目标

通过"化工专业英语"课程的学习，具备化工类中外资企业员工所需达到的英语应用能力的要求，能借助英语开展化工类专业领域的口语交流、工作过程应用文起草、专业文献阅读与翻译工作，提升英语在化工领域的专门化应用能力，巩固和提高化工专业能力，运用英语开展化工专业实践工作和促进职业生涯的可持续发展，较好地服务化工专业实践工作。"化工专业英语"课程学习三角模型如 Fig.1-1 所示。

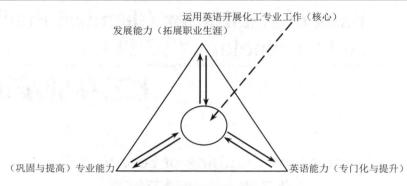

Fig.1-1 "化工专业英语"课程学习三角模型

2．具体目标

① 能与来化工企业合作的外方人员进行交流，并用英语介绍化工企业典型生产的产品、生产工艺、装置等（譬如，年产 100t 的均苯四甲酸二酐产品——pyromellitic dianhydride，PMDA，及其氧化、水解、精制生产工段）。

② 能运用化工专业英语常见词汇、语法、句型，汇报专业实践工作。

③ 能借助英汉化学化工字典、英汉翻译软件等工具，阅读和翻译国外英文化工专业资料（如：生产工艺流程说明、操作说明、设备说明书、维修指南、专利、企业标准）；能将国内先进的化工生产工艺、产品等以英文的形式介绍给外方。

④ 能运用 300 个以上化工专业英语技术词汇（technical words），500 个以上化工专业英语次技术词汇（sub-technical words）撰写化工专业工作过程的检验报告、安全生产记录、合同、产品简介等工作应用文。

3．其他目标

通过"化工专业英语"课程的学习，提高自我学习（active study）能力、计算机（computer）应用能力、化工文献资料检索（retrieval）能力以及与他人沟通协作(cooperation)能力。

Lesson Two　Characteristics of ECET
化工专业英语特点

1．化工专业英语具有专业性强、信息量大的表达特点

化工专业是针对化工过程的普遍规律的描述，不关注个人的心理情绪，具有很强的客观性（objectivity）、无人称性（impersonality）。化工专业英语长句多、被动语态（passive voice）使用频繁，常用"It is …"句型，专业术语（technical words）、缩略词（acronym）经常出现，插图、公式、数字与合成新词所占比例大。例如：

It should be made clear that the cumene route for the production of acetone is more complex but yields another important petrochemical as well as acetone and that benzene is alkylated with propylene in the reaction in the same time.

$$C_6H_6 + CH_2\!\!=\!\!CHCH_3 \longrightarrow C_6H_5CH(CH_3)_2$$
benzene　　　　propylene　　　　　　cumene

以上有关石油化工专业英语的一个长句，包含了219个英文印刷符号和9个化工专业英语技术词汇（benzene 苯、propylene 丙烯、cumene 异丙基苯、acetone 丙酮、complex 混合物、route 工艺流程、petrochemical 石油化工、alkylate 烷基化、reaction 化学反应），句子采用"It should be… that…"句式和被动语态"is alkylated"。该句句子长、专业术语多、化学反应式的出现、被动语态的使用，这些都充分体现了化工专业英语专业性强、信息量大的表达特点。

2．化工专业英语具有客观性、精练性与准确性的语法特点

（1）客观性（objectivity）　因为化工专业是针对化工过程的普遍规律的描述，化工专业英语所要表达的主要内容必定是客观性的，对应于语法则经常采用强调动作的承受者的被动语态，以及表达客观规律的一般现在时态等。

（2）精练性（conciseness）　既然上面已提到化工专业英语句子长、信息量大，为什么还要说在语法的表现形态上具有精练性的特点？可以这样理解，正是因为化工专业的深奥与技术的复杂，造成表达上的信息量大、句子长，倘若语法表达不够精练的话则会导致化工专业英语句子更复杂。从此角度考虑，化工专业英语的语法必须是精练的。

（3）准确性（accuracy）　化工专业英语语法的准确性主要决定于化工专业技术与知识本身内涵。譬如，某个特定的反应温度、反应压力是不能加以随意修饰或者更改的。在化工专业英语中很少看到"It was said…"（据说）句式，这种表达不确定的内容在化工专业中是不可以出现的。相反，数字、公式、插图、表格与反应式等准确性表达方式则经常使用。

3．化工专业英语具有以专业术语多、缩略词语多、合成新词多为主要的用词特点

化工专业的专业性决定了化工专业英语用词上必然大量采用化工领域的专业术语，在化工领域大家熟知的内容，没有必要使用完整的表达形式，因此，缩略现象在化工专业英语中随处可见，譬如实验室"laboratory"通常缩写为"lab"，催化剂"catalyst"通常缩写成"cat"。随着科技的迅猛发展，化工专业技术领域的新概念、新理论、新工艺与新产品不断出现，为此，就会不断出现一些新的专业词汇，例如，当时出现的维他命"vitamin"，也就是后来为大家所认识的维生素。

概括起来，化工专业英语的词汇主要有以下9个特点：专业术语多；次专业技术词汇使用频率高；缩略语使用频繁；合成新词多；非限定动词使用频率高；名词短语多；特用词经常被使用；功能词（介词、冠词与副词等）多；希腊语词根与拉丁语词根比例大。

Lesson Three　Vocabulary of ECET
化工专业英语词汇

1．化工专业英语词汇的种类

（1）技术词汇（technical words）

技术词汇是指适用于某一技术领域使用的专业术语，因而其专业性很强。在化工专业技

术范围内，有很多化工技术词汇。譬如：distillation（分离），reactor（反应器），heat exchanger（换热器），component（成分），catalyst（催化剂），polyethylene（聚乙烯）等。

（2）次技术词汇（sub-technical words）

次技术词汇是指在各个专业领域都较常使用的技术词汇，有时在各个不同的技术领域其涵义会有所区别。例如：temperature（温度）、pressure（压力）、volume（体积）在化工、机械、自动化等专业中都经常使用。composition 和 level 在化工领域中分别是成分与液位的概念，而在日常生活中这两者分别是作文和水平的意思。

（3）特用词（specific words）

在普通英语中，为了使语言生动活泼，经常使用一些短小的词或词组。而在专业英语中为了客观、准确、严谨以及不引起歧义却往往选用一些较长的特用词。

譬如，为了说明电灯亮了，在普通英语中常用：

The light is turned on.

而在专业英语中，常常表述为电灯的电路接通了：

The circuit is completed.

这是因为 complete 词义单一准确，不会产生歧义。而 turn on 不仅可以表示接通，还可以表示其他意义。例如：

We suggest we do not turn on（依赖） our parents.

（4）功能词（function words）

功能词主要包括介词、副词、连词、代词与冠词等，它是单个句子结构以及句子与句子结构之间不可缺少的语言单位。功能词对于理解专业英语的内容十分必要，而且出现频率极高。据统计表明，在专业英语中出现频率最高的 8 个功能词依次是：the,of,in,and,to,is,that,for,are,be。下例化工专业英语句子，其 18 个词中功能词就占了 10 个。

Heat is only one form of energy, though it is certainly the most common and hence most important.

2. 化工专业英语词汇的构成

词汇的形成（word building）：

一是词汇的合成（composition），例如：waterproof（防水的）来源于 water 水与 proof 防止的组合。又譬如，low-boiling-point（低泡点），workshop（work + shop，车间），compressed-air（压缩空气），high-strength（高强度）等。

二是词汇的转换（conversion），例如：coat 既可以当名词"涂层"又可以作为动词"涂覆"。

三是词汇的派生（derivation），例如：名词 nature 加上后缀-al 成为形容词 natural；形容词 saturated（饱和的）加上前缀 un-就成为 unsaturated（不饱和的）。

在上述三种形成专业英语词汇的方法中，派生法（derivation）是特别值得注意的。专业英语词汇尤其是专业词汇大部分是通过派生法形成的，它是在已有的词汇上，通过在词根上加上不同的前缀形成词性不变而词义发生改变的新词，以及通过在词根上加上不同的后缀改变词性，有时也改变词义的新词。由派生法形成的专业词汇通常是较长的，一般的英语词典甚至英汉化学化工词典都无法查寻，只有根据词汇的派生规律去进行推理，结合上下文才能理解词义。为此大家在"化工专业英语"的课程学习中，注意把握专业词汇的派生规律，尤其要掌握常用的化学化工词根、前缀与后缀，否则化工专业技术词汇的学习将是一件十分困难的事情。这里列出一些常用的化工词汇前缀与后缀,供大家参考。

前缀	含义	前缀	含义
a-,an-	非,无,异,不	multi-	多
anti-	抗,反,防	non-	非,不
auto-	自动	under-	不足,在……之下
bi-	二,双	octa-	八
by-	副,侧	ortho-,*o*-	正,原,邻(位)
centi-	百分之一	over-	超,太,过
cis-	顺	para-,*p*-	对(位)
co-,col-,com-,con-	同,共,联	penta-	五
counter-	逆,反	phono-	声
di-	单,一	photo-	光
dis-	分开,无,不	poly-	聚,多
do-	脱,除,反	pre-	预先,以前
extra-	超出,在……之外	re-	反对,再,重新
hepta-	七	semi-	半
hetero-	异,杂	sub-	次,亚,下
hex-	六	sur-	上,超
homo-	同	tele-	遥,远
hydro-	氢的,水的	tetr-	四
hyper-	超	*trans*-	反
hypo-	低,次,下	tri-	三
in-,im-,il-,ir-	不,非	ultra-	超
infra-	亚,下	un-	非,无,不
inter-	相互,在……之间	under-	次,低
iso-,*i*-	异	uni-	单
kilo-	千	-ane	烷
m-,meta-	间(位)	-ate	酸盐
micro-	微,小	-ene	烯
mid-	中	-one	酮
milli-	千分之一	-ose	糖
mini-	小	-ster	酯
mono-	单		

3. 化工专业英语词汇的缩略

词汇缩略是将较长的单词取其首部或主干构成与原词同义的短单词。或者将组成短评的各个单词的首字母拼接为一个大写字母的字符串。在某一个专业领域内,对于大家皆知的事物,或者为了方便起见,专业英语中经常使用缩略词。同时缩略词的使用也方便了印刷、书写、速记与交流。现在在专业英语文献中缩略词的使用越来越频繁。

(1) 节略词(clipped words)

节略词是为了方便起见,用单词的前几个字母形成一个表达同样含义的新词。譬如:

laboratory —— lab 实验室

catalyst —— cat 催化剂

university —— uni 大学

advertisement — ad 广告

（2）缩略词（acronym）

缩略词是用词组的首字母组成的新词，发单词音（不发字母音）。例如：

read only memory — ROM 只读处理器

random access memory — RAM 随机处理器

radio detecting and ranging — RADAR 雷达

（3）首字词（initials）

首字词也是用词组的首字母组成的新词，但发字母音（不发单词音）。例如：

liquid petroleum gas — LPG 液化石油气

gross domestic product — GDP 国内生产总值

gross national product — GNP 国民生产总值

（4）缩写词（abbreviation）

缩写词是由单词变化而来，而且大部分缩写词末尾或每个字母后都有一个句点。譬如：

for example — e.g. 例如

figure — fig. 图，数字

namely — viz. 即，同样是

Lesson Four　Writing and Speaking Frequently Occurred in ECET
化工专业英语常用写作与会话

　　虽然化工专业英语在词汇、语法等方面和普通英语有诸多相同之处，但由于其在科技领域的长期使用，逐渐形成了自身的一些特点。化工专业英语写作是科技英语写作的一个类别，是运用英语开展化工专业工作交流和从事化工科研工作的重要手段。

1. 化工专业英语写作的词汇特点

　　（1）词义专一　为了准确地反映自然界的客观规律，并进行探讨和研究，专业英语的用词要求词义明确专一，避免词义模糊或一词多义的现象。例如：在专业英语中常使用 speculate 来代替日常使用的单词 consider，使用 exceed 来代替日常使用的单词 go beyond，使用 respiration 来代替日常使用的单词 breath，使用 collide 来代替日常使用的短语 run into one another，使用 circulate 来代替日常使用的单词 circle，使用 supervise 来代替日常使用的短语 watch over，使用 mobile 来代替日常使用的单词 movable，使用 synthetic 来代替日常使用的单词 man-made，使用 aviation 来代替日常使用的单词 flying，使用 illuminate 来代替日常使用的短语 light up，使用 edible 来代替日常使用的单词 eatable，以及使用 decompose 来代替日常使用的短语 go to pieces 等。了解并掌握专业英语写作的用词特点，有助于在化工专业英语写作时用词规范，表达地道和清晰。

　　（2）化工专业英语大量使用名词化结构（nominalization）的写作特点　名词化结构是指大量使用名词和名词词组，即在普通英语中使用动词、形容词等充当某种语法成分，而在专业英语中通常转化为名词和名词词组充当语法成分。因为科技文体要求行文简洁、表达客观、内容确切、信息量大、强调存在的事实，而名词正是表物的词汇，因此专业英语惯用名词来表达。此外，名词化结构是以短语形式来表达的，句子言简意赅，可以将更多的信息结构融于一体。例如 Archimeds first discovered the principle of displacement of water by solid bodies. （阿基米德最先发现固体排水的原理。）句中 of displacement of water by solid bodies 系名词化结构，该用法一方面简化了同位语从句，另一方面强调了 displacement 这一事实。例如 If you use firebricks round the walls of the boiler, the heat loss can be considerably reduced.（炉壁采用耐火砖可大大降低热耗。）句中用 heat loss 表示热耗。又如一般文体：When we had completed the experiment, we immediately recorded the result.专业英语文体：On completion of the experiment, we immediately recorded the result. 在专业英语文体中用名词 completion 来表达普通英语中动词 complete 的含义。

2. 化工英语写作的语法特点

化工英语文章的句子结构较普通英语的句子结构更为复杂，且长句较多。这是因为长句更为周密细腻，包容量大，有利于表达复杂意思和更精确地揭示事物间的内在联系。例如下列对高铝红柱石材料的介绍。

Although high density and pure mullite（高铝红柱石） materials have been obtained from small laboratory batches（一次操作需要的原料量）of Al_2O_3-SiO_2 gel, using hot pressing as a consolidation（凝固） route, little attention has been paid in the literature to evaluate the parameters that control the overall processing of mullite gels.

句子长，信息量大，充分体现了化工专业英语的特点，对化工专业知识了解甚少的人，阅读起来很困难。

由此可见，要掌握化工专业英语的写作技巧，必须具备较好的英语基础和造句能力。支离破碎的句子难以表达缜密的思想，也会妨碍信息的准确交流。另外，从上面的例子中亦可看出，为比较客观地进行描述和讨论，避免主观武断，化工专业英语中的被动语态用得尤为广泛。这是因为科技文献侧重叙事推理，强调客观准确。第一、二人称使用过多，会造成主观臆断的印象。因此尽量使用第三人称叙述，采用被动语态。此外，科技文章中将主要信息前置，放在主语部分，这也是广泛使用被动态的主要原因。例如，下面是关于水银测温仪的介绍。

For measuring temperatures below $-40°F$，thermometers filled with alcohol are used. Because of the low melting point of glass, it is limited to use these thermometers for temperatures as high as 1,000°F, as the rising of mercury and the increase of the gas pressure.

这两个长句使用了被动语态和一般现在时，这在化工专业英语中极为常见。一方面被动语态句子允许将最重要的信息放在句首，比主动语态句子更直接明了；此外，工程专业技术人员关心的是专业事实及行为而不是行为者，上例中并未指出谁用水银测温仪测量，谁限定了测温仪的极限，采用被动语态就避免了这种不必要的考证。正因为如此，被动语态广泛地应用于描述专业原理、过程等。就时态而言，化工专业所涉及的内容一般都没有特定的时间关系，所以常采用一般现在时，进一步突出了专业的客观性，本句亦是如此。

3. 化工英语写作的文体结构特点

（1）描述要求具体、准确　例如：Winds between 15 and 30 mph, when accompanied by snow and temperature between 10 °F and 30 °F, often create unstable slabs in avalanche（坍方）-starting zones.

（2）用词造句力求简洁、明了　例如：A series of runs made under identical conditions often yielded different results.

（3）运用图表、公式、符号、缩写词语　运用图表、公式、符号、缩写词语等来替代和简化文字描述，使论述和说明更为直观和简洁。

（4）较多地使用各类复合词结构　例如，linear-expansion, metal-cutting machine, fine-grained steel, light-tight material（防光材料）, moderator-reflector（减速反射器）。

（5）使用通用的固定格式　化工英语写作根据内容的不同往往有一些固定的格式和要求，如论文、说明书、实验报告、信函等。掌握一些通用的固定格式，对写作有很大的帮助。

（6）化工专业文体要求行文简练，结构紧凑　为此，往往使用分词短语代替定语从句或状语从句；使用分词独立结构代替状语从句或并列分句；使用不定式短语代替各种从句；使用动名词短语代替定语从句。这样可缩短句子，又比较醒目。试比较"热量由地球辐射出来时，使得气流上升。"这句话的两种英文表达方式。一般文体：When heat radiates from the earth, it causes air currents to rise. 化工专业英语文体：Radiating from the earth, heat causes air currents to rise.

4．化工专业英语论文写作常用句子

在使用英语写作化工论文的时候，我们要特别注意措辞造句的技巧，以免造成歧义、误解或者表达不清楚的情况。以下是在使用英语写作论文时，常用的一些句式，记住这些句式，对于在专业论文写作时的正确表达有很大益处。

Abstract　摘要

① A basic problem in the design of… is presented by the choice of a … rate for the measurement of experiment variables.

② This paper examines a new measure of…in… based on fuzzy mathematics, which overcomes the difficulties found in the … measures.

③ The method involves the construction of… from fuzzy relations.

④ This paper describes a system for the analysis of the …

⑤ The brief methodology used in… is discussed.

⑥ The technique used is to employ a newly developed…

⑦ The usefulness of… is also considered.

⑧ The procedure is useful in analyzing…

Beginning　开头

① In this paper, we focus on the need for…

② This paper proceeds as follow…

③ The structure of the paper is as follows.

④ In this paper, we shall first briefly introduce fuzzy sets and related concepts.

⑤ To begin with we will provide a brief background on the…

Introduction　内容介绍

① This will be followed by a description of the fuzzy nature of the problem and a detailed presentation of how the required membership functions are defined.

② Details on ... and ... are discussed in later sections.

③ In the next section, after a statement of the basic problem, various situations involving possibility knowledge are investigated: firstly, an entirely possibility model is proposed; then the cases of a fuzzy service time with stochastic arrivals and non fuzzy service rule is studied; lastly, fuzzy service rule are considered.

Review 综述

① A brief summary of some of the relevant concepts in... and... is presented in this article.

② In the next section, a brief review of the.... is given.

③ In the next section, a short review of... is given with special regard to...

④ Section 2 reviews relevant research related to....

Body 主体

① Section 1 defines the notion of robustness, and argues for its importance.

② Section 1 devoted to the basic aspects of the FLC decision making logic.

③ Section 2 gives the background of the problem which includes ...

④ Section 2 discusses some problems with and approaches to, natural language understanding.

⑤ Section 2 explains how flexibility which often ... can be expressed in terms of fuzzy time window.

⑥ Section 3 discusses the aspects of fuzzy set theory that are used in the...

⑦ Section 3 describes the system itself in a general way, including the ... and also discusses how to evaluate system performance.

⑧ Section 3 describes a new measure of...

⑨ Section 3 demonstrates the use of fuzzy possibility theory in the analysis of...

⑩ Section 3 is a fine description of fuzzy formulation of human decision.

Summary 小结

① This paper concludes with a discussion of future research consideration.

② Section 5 summarizes the results of this investigation.

③ Section 5 gives the conclusions and future directions of research.

④ Section 7 provides a summary and a discussion of some extensions of the paper.

Introduction of Time 时间介绍

① Over the course of the past 30 years, ... has emerged form intuitive.

② Technological revolutions have recently hit the industrial world.

③ The development of ... is explored.

④ During the past decade, the theory of fuzzy sets has developed in a variety of directions.

⑤ The concept of... was investigated quite intensively in recent years.

Objective / Goal / Purpose 目标

① The purpose of the inference engine can be outlined as follows...

② The ultimate goal of the ... system is to allow the non experts to utilize the existing knowledge in the area of manual handling of loads, and to provide intelligent, computer aided instruction for...

③ The paper concerns the development of a...

④ The scope of this research lies in...

⑤ The main theme of the paper is the application of rule based decision making.

⑥ These objectives are to be met with such thoroughness and confidence as to permit...

⑦ The objectives of the ... operations study are as follows...

⑧ The primary purpose/consideration/objective of...

Problem / Issue / Question 问题

① Unfortunately, real-world engineering problems such as manufacturing planning do not fit well with this narrowly defined model. They tend to span broad activities and require consideration of multiple aspects.

② Remedy / Solve / Alleviate these problems...

③ What is a difficult problem, yet to be adequately resolved?

④ Two major problems have yet to be addressed.

⑤ This problem in essence involves using x to obtain a solution.

⑥ An additional research issue to be tackled is...

⑦ Some important issues in developing a... the system is discussed.

⑧ The three prime issues can be summarized...

5. 化工专业英语常用会话注意事项

我们在化工专业实践过程中除需要书面交流以外，很多时候我们还需要口语交流。在口语交流时，首先要掌握常用的日常英语口语；其次，注重积累化工专业词汇和次专业词汇，譬如，"composition"在普通英语中是作文之意而在化工专业英语中是"成分"的含义，同样"level"在化工上表示"液位"，"critical"是指"临界（温度）"等；再次，还应该注意英语口头交流的一些基本技巧和其他有关文化方面的问题，强化基本的英语口语技巧以及化工专业实践口语训练。

第一，多学习、多交流、多积累。注意化工专业英语的有效单词量和习语积累。积累中不仅要记住单词的读音、拼写、中文语义，更重要的是它所使用的语境、用法、传达的信息等。同时不要单纯地死背单词，一定要放在上下文语境中记忆，能够让人事半功倍，在这方面可以通过多查阅国外化工类英文网站进行训练。平时多说英语，找个和你练习英语的同伴，练练自己的口语。平常不要放过和外国人聊天的机会，毕竟学口语都是为了和人交流的，所以"说"很重要。多看些外国电影或者多听英文录音，练习自己的听力，也可以尽量模仿电影里的语调，练习自己的发音。

第二，灵活运用多渠道表达的方法。人与人之间交谈80%是想告诉对方这个事物是什么。如果一种表达式对方不懂，可以寻找另一种表达式最终让对方明白。因为事物就一个，但表达它的语言符号可能会很多。也就是说用一种不同的方式表达同一个意思，或者一个表达式对方听不清楚，举一个简单易懂的例子来表达，直到对方明白。例如，你想告诉对方你今天中午吃了什么东西，但是不知道中午吃的这些菜品如何翻译为英文（对于中餐来说这种情况非常常见），那你完全可以通过描述这道菜用了哪些材料，如何制作，通过简短的描述就能让对方理解你的意思。

第三，不要怕犯错，尽量多开口交流。只要不是在法庭上这种一句话都不能说错的场合，日常的口语交流犯错没有什么大不了的。很多同学日常说话是怕犯错，怕听不懂，怕用错语法，其实大可不必。即使是母语是英语的人，平时说话也难免犯些小错误，平时互相交流有时一两句话没听清也是常事，不必过度在意。没听清对方的说话，只要礼貌地请对方重复一遍就行。一些小的不影响表达的小语法错误也不会让对方听不懂你的说话。只要大着胆子，

Unit One　Basis of English for Chemical Engineering and Technology (ECET)

多说，多交流，假以时日，口语能力一定会进步。

　　第四，以下为化工专业实践与日常生活常用的近百句口语，供大家参考。

　　※On behalf of my company, I will introduce this project. 请允许我代表公司，介绍一下该项目。

　　※What do you think of this chemical technique? 你觉得这项化工技术怎么样？

　　※See to this device maintenance. 确保设备维护到位。

　　※See to it that our team punches in at 8:00 every day. 确保班组每天 8:00 准时上班。

　　※I don't agree to your project more. 我完全同意你提出的项目。

　　※I don't agree with you any more. 我不太同意你的观点。

　　※This mixture is made up of three components. 这个混合物是 3 种成分组成的。

　　※The level of this reactor is 1.2 meters. 此反应器的液位是 1.2m。

　　※The temperature of this distillation column is 150.5℃. 该精馏塔的温度是 150.5℃。（℃: degrees Centigrade；℉:degrees Fahrenheit）

　　※The raw materials will run out. 原料快要用完了。

　　※What's the point of this unit operation? 这项化工单元操作的重点是什么？

　　※I am in charge of the technical side. 我主要负责技术上的事。

　　※Please change the catalyst. 请更换催化剂。

　　※Let's go to the heat exchanger. 我们去换热器那儿看看。

　　※We must observe safety rules of our company. 我们必须遵守公司的安全规程。

　　※It's just one of those things in the chemical process. 这是化工过程中常有的事。

　　※Xiao Zhang is our shift manager. 小张是我们的值班经理。

　　※Don't forget to punch in. 别忘记做上班考勤记录。

　　※We are going to go for practice next week. 下星期我们要去实习了。

　　※That will widen my experience and horizon of employment. 这将拓展我的职业经验和发展空间。

　　※Sorry, I have to ask for leave. 抱歉，我得请假。

　　※We are preparing to go off work. 我们正准备下班。

　　※What time do we work overtime today? 今天我们何时加班？

　　※I am interested in the job you advertised in the local newspaper. 我对你们刊登在当地报纸上招聘的那份工作很感兴趣。

　　※Is there an opening in your department? 你们部门还有岗位空缺吗？

　　※Why do you try our corporation? 你为什么到我们公司来就业？

　　※Are you prepared to do probationary rear? 你准备好做后方见习了吗？

　　※I majored in the chemical engineering. 我主修化学工程专业。

　　※I beg your pardon. 请你再说一遍。

　　※I see. 我明白了。

　　※Thank you very much for your help in advance and I am looking forward to hearing from you soon. 提前对你的帮助表示感谢并期盼回复。

　　※Come on! 赶快行动！

　　※Work it out. 难题解决了。

※Did you come up with any ideas about this new product? 对新产品有没有其他建议？

※Ladies and Gentlemen, all distinguished guests present here… 女士们、先生们，尊敬的各位来宾……。

※Stay up. 熬夜工作。

※Up to one's eyes. 忙于……。

※Take it easy. 别紧张。

※Take your time. 不着急。

※Here you are. 给你。

※How do you do! 你好！（用于陌生人之间）

※How are you? 你好！（用于熟人之间）

※Please pass me your resume. 请将你的简历给我。

※By the way. 顺便说一下。

※It does. 某事行。

※I can manage it. 我能行。

※This is…这里是……。（电话用语）

※What's the date today? 今天是几号？

※What day is it today? 今天星期几？

※Do you have the time? 现在几点钟？

※Cheers! 干杯！（喝一口）

※Bottom up! 干杯！（喝完）

※Can I help you? 我能为您效劳吗？

※Go Dutch. 各自付账。

※Keep the changes. 不用找零。

※Hold on a minute./ Hold the line. 别挂机，请稍等。

※You're welcome./ Don't mention it./ It's my pleasure. 不客气。

※It's a piece of cake./No problem. 没问题。

※Is this seat taken? 这儿可以坐吗？

※Please be seated. 请坐。

※It's only between you and me. 请保密。

※Your remarks are beyond the reach of me. 我没听懂你说的话。（委婉语）

※Figure out. 我明白了。

※Lay off. 解雇。

※Make up. 弥补损失。

※The other day/ the day after tomorrow. 几天前/后天。

※Drop me a line! 写封信给我。

※Pull over! 请把车子开到旁边。

※What a big hassle. 真是个麻烦事。

※Go for it. 加油。

※So do I. 我也一样。

※I am racking my brains. 我正在绞尽脑汁。

※May I take a rain check? 可不可以改到下次？
※Take a back seat. 大家都退一步。
※Green hand. 生手、没有经验的人。
※This is in way over my head. 对我而言这实在太难了。

Lesson Five Grammar Frequently Used in ECET
化工专业英语常用语法

1. 化工专业英语五种句式

（1）SV(SV_i)——主谓

　　We will do.

（2）SVO(SV_tO)——主谓宾

We are learning English for Chemical Engineering and Technology.

（3）SLP（SLP）——主系表

Fifty students are to go for practice.

（4）SVOO($SV_tO_iO_d$)——主谓双宾

Mr John has sent me an E-mail.

（5）SVOC(SV_tOC)——主谓宾补

Engineer Zhang asked us to carry out the chemical experiment of EDTA. 其中，

S——subject 主语

V——verb 谓语动词

V_i——intransitive verb 不及物动词（无被动语态并且不能带宾语）

V_t——transitive verb 及物动词（可以有被动语态并且必须带宾语）

L——link verb 系动词（包括较常用的系词 be）

P——predicate 表语

O——object 宾语

O_d——direct object 直接宾语

O_i——indirect object 间接宾语

C——complement 补语

2. 化工专业英语中句子之间的连接

（1）Sentence One and Sentence Two（两个独立主句用 and 连接）

We are drinking coffee and they are drinking Pu'er tea.

（2）Sentence One ; Sentence Two（两个独立主句用分号连接）

We are learning common English; we are to learn English for Chemical Engineering and Technology next semester.

（3）Sentence One; therefore Sentence Two（两个独立主句用分号连接并加上连接副词的拓展形式）

Xiao Wang did his best; therefore he passed the terminal examination last term.

（4）Sentence One. Sentence Two（两个独立主句用句号连接）

Both of us are fresh boys. Zhang xiao is a sophomore.

（5）Sentence One. Moreover Sentence Two（两个独立主句用句号连接并加上连接副词的拓展形式）

The city's museums are great buildings. Moreover, they are cultural centers.

此外，我们要特别注意，在英文中两个独立主句是绝对不可以用逗号连接起来。或许，有些朋友要问，我们经常在英文阅读中碰到逗号，那是怎么回事？的确，逗号在英文中使用还是比较频繁的，它主要适用于并列成分相连、从句与主句相连、非谓语动词与主句相连、状语与主句相连等情况，但不可以是两个独立的主句用逗号连接。譬如：

If you come here, you will have your idea.（从句与主句相连）

It being hot, our shift manager asked us to stop working yesterday.（非谓语动词与主句相连，该句中如果将"being"换成"was"，就会出现逗号连接两个独立主句的错误情形，整个句子就错了。）

With the tools in our hands, we can start to repair the heat exchanger.（状语与主句相连）

3．主谓一致问题

（1）A and B，谓语动词的单复数取决于 A 和 B 整体。

You and Xiao Li have made good achivements in operating the software of chemical simulated system.

（2）A with B，谓语动词的单复数取决于 A（with B 是介词短语，所以不可以作主语）

John with other workers doesn't want to go picnicking this week.

（3）A as well as B，谓语动词的单复数取决于 A（as well as 短语，其语义重心在前面）

The professional morality as well as knowledge is very important.

（4）not only A but also B，谓语动词的单复数取决于 B

Not only three managers but also a worker is to have a chance to go abroad.

（5）either A or B，谓语动词的单复数取决于 B（就近规则）

Either you or he is made to send for a doctor.

（6）neither A nor B，谓语动词的单复数取决于 B（就近规则）

Neither you nor Tom hasn't passed the final examination.

（7）There is/are A and B. 谓语动词"is/are"的单复数取决于 A（就近规则）

There is a book and two pens on the desk.

4．名词性从句（noun clause）

（1）主语从句（subject clause）

It is … that/what/how/when/whether

It is evident that mechanical or external pressure must be used in order to transfer the oil.

（2）宾语从句（object clause）

The process requires that the bottoms from the Tower 101 be moved to the top of Tower 102.

（3）表语从句（predicative clause）

Common to all the techniques is that no change of molecular structure occurs during the operations and no new compounds are formed.

（4）同位语从句（appositive clause）

同位语从句主要用于解释说明一些抽象名词的含义和内容。常见的抽象名词有：possibility, belief, idea, evidence, fact, thought, new 等。引导词有 that, what 和 how。在化工专业英语中为了解释一些术语，经常会用同位语从句进行补充说明。

A main distinguishing feature of the various petroleum products is their volatility that they have the ability to vaporise.

I have no idea what he was referring to.

The fact that he has already done his best seems to have been ignored.

5. 定语从句(attributive clause)

又称关系从句（relative clause）、形容词性从句(adjective clause)，在化工专业中为了准确阐明某一专业行为，经常使用定语从句。

（1）结构

This is the house which Lu Xun ever lived in.

先行词（antecedent）——the house

关系词，这里是关系代词（relative pronoun）——which。关系词包括：①关系代词，that, which, who, whom, whose, as 等；②关系副词，when, where, why 等。关系词一定要在定语从句中充当一定的成分，这是判断是否为定语从句的核心环节。把握定语从句要注意其"3 个要素"与"1 个核心"。"3 个要素"是指先行词、关系词和从句，它是定语从句的表征；"1 个核心"就是指关系代词或者关系副词必须在从句中充当某一成分，它是定语从句的本质体现。否则 that 之类的词就不是关系词而是连词，仅起连接作用，不作成分。

从句（clause）——Lu Xun ever lived in.（lived in 的宾语为前面的 the house）

（2）类别　限制性定语从句（restrictive attributive clause）表示对某一事物具有明确的限定性；非限制性定语从句（non-restrictive attributive clause）强调惟一性、整体性，其标志是在关系词前有逗号。

下面通过两个句子来具体体现这两者的区别。

The travellers who knew about the floods took another road.（知道洪水泛滥的旅客都改道而行了。）限制性定语从句，意指不知道洪水泛滥的部分旅客仍然走原路。

The travellers, who knew about the floods, took another road.（旅客们知道洪水泛滥，都改道而行了。）非限制性定语从句，体现整体性、惟一性，意指所有旅客都知道洪水泛滥，都改道而行了。

（3）定语从句相关例句

This is the best specimen that we have got.

The very book that I lost is very significant to me.

His mother always gives him more money than he needs.

Here are such questions as are frequently asked by freshmen.

We don't know the reactor for which he is looking.

All he knows is that we will go for practice.

This is the second experiment that I conducted today.

This is the house where Lu Xun ever lived.

This is the reason why the football team lost the game last week.

The reason which he has given for not coming is that he caught a cold yesterday.

Is there anything I can do for you?

As we know, China is a great country.

China is a great country, which is known to all.

The nature of the crude decides to a certain extent the nature of the products that can be manufactured from it and their suitability for special application.

6. 状语从句（adverbial clause）

（1）时间状语从句(adverbial clause of time) 引导时间状语从句的连词主要有 when, while, as, once, since, before, after, until, as soon as, hardly… when，no sooner… than 等。有些副词和名词也可用作从属连词，引导时间状语从句。譬如，instantly, immediately, directly, the instant, the day, the minute, the moment, the second 等。

When the vapour pressure is equal to or slightly higher than atmospheric pressure, vapour forms freely throughout the whole liquid.

As distillation proceeds, the composition of both distillate and residue will change progressively until all the liquid has been distilled into the receiver.

The instant our shift manager came in, we started to carry out our experiment.

（2）地点状语从句(adverbial clause of place)

Go where you are needed by our motherland.

You had better put the tools where they can be seen.

（3）原因状语从句(adverbial clause of reason) 引导原因状语从句的连词主要有：because, as, for, since, now (that), in that, considering that 等。

Ethylene glycol（乙二醇）was a landmark of probably even greater significance than the production of IPA（Isopropyl alcohol 异丙醇）, in that it was the first petrochemical derived from ethylene.

Theory is valuable because it can provide a direction for practice.

（4）方式状语从句(adverbial clause of manner) 引导方式状语从句的从属连词有 as, as if(though), the way, how 等。

He made some changes as you had suggested.

He speaks English fluently as if he were an Englishman.

（5）让步状语从句(adverbial clause of concession) 引导让步状语从句的从属连词有 no matter（what, how, where, when）, though, although, while, however/whatever/wherever/whenever, as, even if（though）, whether… or 等。

Wherever you go and whatever you do, I'll be waiting for you.

While we don't agree, we continue to be friends.

Tired as he may be, he will go to the lecture.

（6）目的状语从句（adverbial clause of purpose） 引导目的状语从句的从属连词有 so that, in order that, in case, lest, for fear that 等。

Care must be taken in using this method lest overflow should occur.

Take an umbrella in case it rains.

（7）结果状语从句（adverbial clause of result） 引导结果状语从句的连词有 so that, so…that, such…that 等。

It was such a warm day that Peter took off his overcoat.

So quickly did he write the answers that he finished the examination in an hour.

（8）比较状语从句（adverbial clause of comparision）

The weather of Wuhan is as hot as that of Nanjing.

I am more a friend of yours than your teacher.

He works twice as hard as anyone else.

（9）条件状语从句（adverbial clause of condition） 引导条件状语从句的从属连词有 if, unless, only if, supposing, suppose, providing, provided, as long as 等。

As long as you persevere you will succeed in the end.

Provided (that) he wins the support of the majority groups, he will be able to win the election.

（10）限制性状语从句(adverbial clause of restriction)

As far as I am concerned, I have no objection to your doing so.

As far as the matter is concerned, I think it possible to succeed.

7．强调句型

强调句型的表达法一般包括以下三种形式。

（1）当句子的主语、宾语、状语需要强调时，强调句型为：It is/was + 被强调部分 + that(who/whom) + 句中其他部分。

It is this property of fluidity that makes pipeline transport of materials possible.

It is the article that they discussed last week.

It was because he was too careless that he forgot to turn off the power.

（2）It is /was not until… that 也为强调句的一种形式，意思是"直到……才"。

It's not until Mr John came to China that he knew what kind of country she's.

（3）do 在句中可用来强调谓语动词，用于一般现在时或一般过去时。

She does like going picnicking.

He did send me a postcard as I expected.

8．比例增减句

比例增减句的表达形式为：the + comparative degree（比较级）…the + comparative degree （比较级），比较级可以是形容词的比较级也可以是副词的比较级。常见形式为：the more … the more…。该句型意指"越……越……"。

The higher the level is, the more pressure will be.

The more work we give our brains, the more work they are able to do.

9．"with 结构"

（1）表达形式

with + O + C （O: object 宾语；C: complement 补语）

具有否定意义的拓展形式：without + O + C

其中补语可以是：①adj 形容词；②adv 副词；③prep 介词；④to do (sth)动词不定式短语；⑤doing（sth）现在分词短语；⑥done 过去分词。

（2）特点　with-structure 是典型的书面语表达形式，结构紧凑，内容简练，因此，在化工专业英语文献资料、英文著作等出现频率较高。

（3）作用

① 作定语，相当于定语从句。

In the corner is a table with one leg shorter than the others.

② 作状语，相当于时间状语从句、原因状语从句、条件状语从句等。

With no paper to write on, she tore a piece of cloth from his shirt.

With the entire job finished, he went to seaside for a holiday.

Manager Wang came into the meeting room with a book in hidden hands.

Only with a new human measures introduced will he regain the control of the company.

Without sun's light warming the earth's surface, no life could exist on the earth.

③ 作伴随状语。

Mr John dislikes sleeping with the light on.

The lady got on the bus with a baby in her hands.

10. 被动语态（passive voice）

语态（voice）是一种动词形式，表示主语与谓语动词所表示的动作之间的关系。英语动词有两种语态：主动语态（active voice）与被动语态（passive voice）。英语被动语态是由"助动词 be + 及物动词的过去分词"构成的。被动语态的时态变化借助系词 be 来实现。助动词 be 要在人称和数上与主语保持一致。在应用当中，当无须指出动作的执行者、动作的执行者不明确时以及强调动作的承受者时，通常使用被动语态。这恰好符合专业英语描述客观事物，具有很强专业性与客观性的特点，为此在专业英语文献中被动语态的使用是十分频繁的，大家注意把握。

（1）专业英语中常用时态的被动语态形式

The fraction of sample entering each branch can be determined by injecting a pure saturated hydrocarbon.（一般现在时）

That heat exchanger was not bought in Shanghai.（一般过去时）

As distillation proceeds, the composition of both distillate and residue will change progressively until all the liquid has been distilled into the receiver.（现在完成时）

Much attention will be focused on this project.（一般将来时）

Two pumps are being repaired by Engineer Li.（现在进行时）

The chemical laboratory building was being built then.（过去进行时）

When he came back, the project problem had been solved.（过去完成时）

（2）常用被动语态的几种情形

① 当关注或者强调动作的承受者时，多用被动语态。这时，因为动作的执行者处于次要地位，句子中 by 引导的短语可以省略。例如，The heat transferred to the liquid in the process of boiling is retained in the vapour(latent heat of evaporation).

② 在专业文献资料中，出现没有必要说出动作的执行者的情形时，通常用被动语态。此时，句子中不出现带有 by 的引导短语。譬如，Reforming processes were developed for the purpose of converting low-octane heavy gasoline fraction (naphthas) into product with a higher ignition quality,in terms of octane number, for blending into motor and aviation gasoline.

11. 动词不定式（infinitive）

动词不定式是非谓语动词的一种形式，它是指带 to 的动词原形（使用时有时省略 to 或者不带 to），在句中具有名词、形容词和副词作用，可以作主语、表语、宾语、定语、状语和补足语。动词不定式时态和语态的变化见 Table 1-1。

Table 1-1 动词不定式时态和语态的变化

时 态	语态（主动）	语态（被动）	用 法
一般时	to do	to be done	表示与谓语动词同时发生或者在其之后发生
进行时	to be doing	—	表示与谓语动词同时发生，强调动作正在进行
完成时	to have done	to have been done	表示动作发生在谓语动词之前
完成进行时	to have been doing	—	表示动作发生在谓语动词之前，强调动作进行状态
综合表达式	（for /of）sb (not)(to)do(sth)	(not)(to)be done	动词不定式有自己的逻辑主语和否定形式

（1）作主语　动词不定式作主语，表示一个特定的行为或事情。按照英语文法的要求，避免头重脚轻，并且遵循英文从抽象到具体的思维原则，常用 it 作形式主语或者形式宾语。

It is essential for us to go for practice in the factory.

We think it significant for us to learn English for chemical engineering and technology.

（2）作宾语　动词不定式在某些及物动词后可作宾语。这类及物动词通常有 advocate, intend, refuse, afford, ask, determine, strive, expect, fail, undertake, promise, begin, wish 等。

They decided to carry out the chemical experiment this afternoon.

He refused to att end the press conference.

（3）作表语　动词不定式位于系动词后作表语，譬如，Our main job this year is to finish the project.

（4）作定语　动词不定式作定语时，通常放在所修饰的名词后面并与其修饰的词之间有动宾关系，如果该不定式是不及物动词，其后必须加上相关的介词。

Women should have the right to receive education.

Could you give me a pen to write with?

（5）作状语

I spoke very slowly so as to make me understood.（作目的状语）

Solar batteries have been used in satellites to produce electricity.（作结果状语）

Mary burst into laughter to see his funny action.（作原因状语）

12. 分词（participle）

分词是动词的三种非谓语动词之一，分词有现在分词（present participle）和过去分词(past participle)。分词主要在句中作状语及定语。现在分词与过去分词的区别主要体现在时间和语态概念上。在时间上，现在分词表示动作正在进行，过去分词则表示动作已经完成。在语态上，现在分词（除被动式外）表示主动意思，过去分词表示被动含义。

（1）现在分词

This is a pressing problem.（作定语）

Climbing to the top of the tower, we saw a magnificent view.（作时间状语）

Being sick, I had to stay at home.(作原因状语)

Adopting this method, we will raise the average yield by 30 percent.（作条件状语）

Admitting what he has said, I still think he hasn't done his best.（作让步状语）

It rained heavily, causing a flood in the suburb.（作结果状语）

Please fill in the form, giving your address and name.（作伴随状语）

The result of the experiment was encouraging.（作表语）

You had better start the computer running.（作补语）

（2）过去分词

Did Tom pay a visit to the tomb of the unknown soldier?（作定语）

When seen through a telescope, the sun appears darker near the edge.（作时间状语）

Given the voltage and current, we can determine the resistance.（作条件状语）

Badly involved in the accident, the car is still running.（作让步状语）

Overcome with surprise, she was unable to utter a word.（作原因状语）

In my office, there is a green typewriter, unrepaired.（作伴随状语）

Some substances remain practically unchanged when they are heated.（作表语）

The teacher spoke so slowly that he wanted to make him understood.（作补语）

13．动名词（gerund）

动名词也是动词的一种非限定形式，由动词原形加词尾-ing 构成。与现在分词的构词方法相同。它没有人称和数的变化，同时具有名词和动词特征，因而又可以将它称为"动词化的名词"和"名词化的动词"，在名中主要充当主语、宾语、表语和补语。

Thermal cracking has mostly been applied for other purpose.（作主语）

You had better take your tools in case some machines require repairing.（作宾语）

The function of a capacitor is storing electricity.（作表语）

French is one of the working languages at international conference.（作定语）

We call this process evaluating.（作补语）

动词不定式、分词与动名词三类非谓语动词在具体应用过程中，应注意把握"三位一体"的要领，它是指"语态、时态与逻辑主语"三个方面形成非谓语动词的完整表达。在实际运用时首先找出其逻辑主语，判定其语态，其次确定其时态，非谓语动词的时态是相对时态，通常发生在主句的谓语动词之前用完成态，其他情形用一般态即可。

14．化工专业英语常用时态(tense)

谓语动词用来表示动作或情况发生时间的各种形式称为时态。在专业英语中常用的时态主要有9种时态（在英语中，一共有16种时态）：一般现在时、现在进行时、现在完成时、一般过去时、过去进行时、过去完成时、一般将来时、将来进行时和将来完成时。

（1）一般现在时　一般现在时通常表示习惯动作、一般状态、客观规律和永恒真理。例如：

She often gets up late on Saturday.（习惯动作）

Professor Li lives in Gulou District Nanjing City.（一般状态）

The sun rises in the east and sets in the west.（客观规律）

During the cracking reactions some heavy material, known as "coke", is deposited on the catalyst.（永恒真理）

（2）现在进行时　首先，现在进行时表示说话时正在进行的动作，也表示目前一段时间内进行的活动。表示后一种情况时，动作也一定正在进行。

They are planning the project of building chemical experiment.

Professor Xia is speaking with him on the phone.

其次，现在进行时也可表示将来的动作：它指按照人们的计划、安排将要发生的动作或者即将开始的动作。

Mr John is leaving for Beijing tomorrow.

Are you seeing someone off?

再次，现在进行时还可以表示长期重复的习惯性动作，带有一定的赞赏或者厌烦的感情色彩。

She is always cooking some delicious food for her family.

It is always raining these days.

（3）现在完成时　现在完成时表示目前已完成的动作或刚刚完成的动作，也可表示从过去某一时刻发生，现在仍延续的动作情况。此时态表示动作发生在过去，但重点强调对现在的影响，是指从现在看未来。

The press conference has lasted for three hours.

Our factory has just bought an unusual heat exchanger.

如果动作仅仅是发生在过去，对现在无影响则用一般过去时，试比较下面两个句子。

What did you sing in yesterday's party?

We have learnt Common English for about 9 years.

（4）一般过去时　一般过去时表示过去某时间发生的动作，也可表示过去习惯性的动作。一般过去时不强调动作对现在的影响，只说明过去。譬如：

How many subjects did you study last semester?

He smoked forty cigarettes a day until he gave up.

As a boy, I used to go fishing at that river.

（5）过去进行时　过去进行时表示过去某时正在发生的动作，也可表示过去某段时间内正在发生的动作或者反复发生的动作,尤其用来描述过去事件发生的背景。

When his teacher came in he was conducting his chemical experiment.

Some students are singing, while others are dancing on the playground.

（6）过去完成时　过去完成时表示过去某时之前已完成的动作或者状态，在时间上，它有明显的发生在过去参照时间之前，属于"过去的过去"，表示动作发生在过去，强调对某一过去时间的影响。

By the end of last term, we had finished our subjects.

Mary said that she had never been to Beijing.

（7）一般将来时　一般将来时表示某个将来时间发生的动作或者状态，也可表示将来反复发生的动作或习惯性的动作。

The students will have six English classes per week next term.

Xiao Li is going to travel around the world.

需要说明的是："will do"表示说话人无计划性的临时想法，这与 will 作名词时的词义"意愿"有一定联系；"be to do"表示按照计划或者正式安排将要发生的事情；"be going to do"强调客观上将要发生。

（8）将来进行时　将来进行时表示将来某时正在发生的持续的动作。其表达形式为：will be +动词的现在分词。例如：

We will be having a mathematics class this time tomorrow.

Most of the young fans will be meeting him at the airport.

（9）将来完成时　将来完成时表示将来某时前完成的动作，它也可以表示推测。例如：

We will have finished ten courses by the end of next term.

Mr Brown will have stayed here for five months next week.

为了便于准确把握英语常用时态，提出英语"点、线、面"时态学习法，具体内容如下，见 Fig. 1-2。

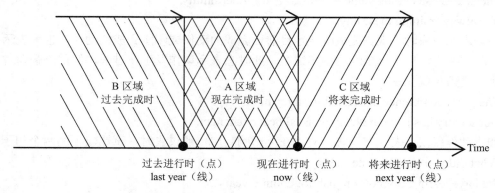

Fig. 1-2　英语"点、线、面"时态学习法示意图

在 time 坐标上任意选取 now, last year, next year 三条与 time 坐标垂直的直线，分别代表一般现在时、一般过去时和一般将来时；此三条直线与 time 坐标的交点表示进行时态，分别代表现在进行时、过去进行时和将来进行时；"A 区域"表示到现在，它指的是动作发生在过去，强调对现在的影响，因此对应现在完成时；"B 区域"表示到上年度，它指的是动作发生在过去，强调对上年度的影响，因此对应过去完成时；"C 区域"表示到明年，它指的是动作发生在将来，强调对明年的影响，因此对应将来完成时，可见完成时态在图上是一个"面积"的概念。归纳一下，大家便能理解："点"代表进行时态（现在进行时、过去进行时和将来进行时），"线"代表一般时态（一般现在时、一般过去时和一般将来时），"面"代表完成时态（现在完成时、过去完成时和将来完成时）。当遇到时态问题时，大家可以画一下"点、线、面"时态图便有助于准确把握时态。

Unit Two

Petrochemical Industry

石油化学工业

Lesson One Petroleum Refinery Distillation
石油炼制

A petroleum refinery is an installation that manufactures finished petroleum products from crude oil, natural gas liquids, and other hydrocarbons. Refined petroleum products include but are not limited to gasolines, kerosene, distillate fuel oils, liquefied petroleum gas, asphalt, lubricating oils, diesel fuels, and residual fuels.

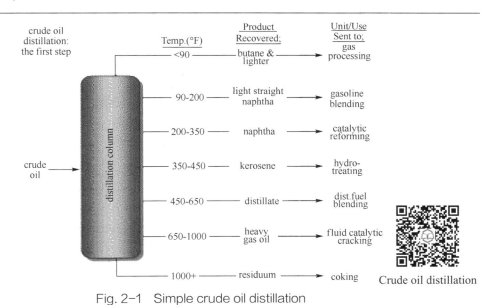

Fig. 2-1 Simple crude oil distillation

The core refining process is simple distillation (Fig. 2-1). Because crude oil is made up of a mixture of hydrocarbons, this first and basic refining process is aimed at separating the crude oil into its "fractions," the broad categories of its component hydrocarbons. Crude oil is heated and put into a still—a distillation column—and different products boil off and can be recovered at different temperatures. The lighter products—liquid petroleum gases (LPG), naphtha, and so-called "straight run" gasoline—are recovered at the lowest temperatures. Middle distillates—jet fuel, kerosene, distillates (such as home heating oil and diesel fuel)—come next. Finally, the heaviest products (residuum or residual fuel oil) are recovered, sometimes at temperatures over 1000 degrees F. The simplest refineries stop at this point. Other refineries reprocess the heavier fractions into lighter products to maximize the output of the most desirable products.

Petroleum industry manufactured products

Additional processing follows crude distillation, "downstream" (or closer to the refinery gate and the consumer) of the distillation process. Downstream processing is grouped together in this discussion, but encompasses a variety of highly-complex units designed for very different upgrading processes. Some change the molecular structure of the input with chemical reactions, some in the presence of a catalyst, and some with thermal reactions.

In general, these processes are designed to take heavy, low-valued feedstock—often itself the output from an earlier process—and change it into lighter, higher-valued output. A catalytic cracker, for instance, uses the gasoil (heavy distillate) output from crude distillation as its feedstock and produces additional finished distillates (heating oil and diesel) and gasoline. Sulfur removal is accomplished in a hydrotreater. A reforming unit produces higher-octane components for gasoline from lower-octane feedstock that was recovered in the distillation process. A coker uses the heaviest output of distillation—the residue or residuum—to produce a lighter feedstock for further processing, as well as petroleum coke.

Words and Expressions

petroleum [pɪˈtrəʊlɪəm]　　　　　　　　n. 石油
refinery [rɪˈfaɪnərɪ]　　　　　　　　　n. 精炼厂，提炼厂，炼油厂
distillation [ˌdɪstɪˈleɪʃən]　　　　　　n. 蒸馏，蒸馏法
hydrocarbon [ˈhaɪdrəʊˈkɑːbən]　　　n. 碳氢化合物
kerosene [ˈkerəsiːn]　　　　　　　　n. 煤油
asphalt [ˈæsfælt]　　　　　　　　　　n. 沥青
lubricating oil　　　　　　　　　　　n. 润滑油
residual [rɪˈzɪdjʊəl]　　　　　　　　adj. 残余的，剩余的
fraction [ˈfrækʃən]　　　　　　　　　n. 馏（部）分
naphtha [ˈnæfθə]　　　　　　　　　　n. 石脑油
feedstock [ˈfiːdstɒk]　　　　　　　　n. （工业加工用的）原料，尤指（用于制造石油化学产品的）化工物
sulfur [ˈsʌlfə]　　　　　　　　　　　n. 硫
octane [ˈɒkteɪn]　　　　　　　　　　n. 辛烷

Notes

1. Because crude oil is made up of a mixture of hydrocarbons, this first and basic refining process is aimed at separating the crude oil into its "fractions", the broad categories of its

component hydrocarbons.

句子分析：句子的结构为"主语+动词的被动语态"，because 引导了一个原因状语从句：因为……所以……。separate into 意为把……分成，分离成；aim at doing sth 意为目的是做……

译文：原油是由烃类组成的混合物，第一步蒸馏的目的是把原油切割成不同的"馏分"——按烃类组成的大致分类。

2. The lighter products—liquid petroleum gases (LPG), naphtha, and so-called "straight run" gasoline—are recovered at the lowest temperatures.

句子分析：句子的结构为"主语+被动语态"，这是一个被动语态的句子：be+过去分词构成。

译文：在最低的温度下得到较轻质产品——液化石油气（LPG）、石脑油和所谓的直馏汽油。

3. Other refineries reprocess the heavier fractions into lighter products to maximize the output of the most desirable products.

句子分析：句子的结构为"主语+谓语+宾语"，to 引导目的状语从句。reprocess into 表示再加工成……

译文：其余精制过程中，较重的馏分转变成较轻产品，以期最有价值产品产量的最大化。

4. Downstream processing is grouped together in this discussion, but encompasses a variety of highly-complex units designed for very different upgrading processes.

句子分析：句子的结构为"主语+被动语态"。这是一个被动语态的句子：be+过去分词构成，but 引导了一个转折的状语从句"但是……"。Encompass 表示包含。

译文：下游加工过程集中一起讨论，但总的说，各种复杂单元设计组合都是为了不同的精制加工目的。

5. A catalytic cracker, for instance, uses the gasoil (heavy distillate) output from crude distillation as its feedstock and produces additional finished distillates (heating oil and diesel) and gasoline.

句子分析：句子的结构为"主语+ 谓语+ 宾语"，and 连接并列的谓语结构。for instance 表示例如。

译文：例如，催化裂化装置是把从常压蒸馏装置得到的粗柴油（重组分）作为原料加工成附加值高的精制馏分油（民用燃料油和柴油）和汽油。

Reading Comprehension

1. Can you describe the refined petroleum products and its intended uses?
2. What are the main compositions of petroleum?
3. What technology is used to refine petroleum?
4. What is the term of "straight run"?
5. What are the reactions in the hydrotreater?

Reading Material

Shale Gas Utilization 页岩气的利用

Shale gas is natural gas that is found trapped within shale formations. Shale gas has become an increasingly important source of natural gas in the United States since the start of this century, and interest has spread to potential gas shales in the rest of the world. In 2000 shale gas provided only 1% of U.S. natural gas production; by 2010 it was over 20% and the U.S. government's Energy Information Administration predicts that by 2035, 46% of the United States' natural gas supply will come from shale gas.

Some analysts expect that shale gas will greatly expand worldwide energy supply. China is estimated to have the world's largest shale gas reserves. A study by the Baker Institute of Public Policy at Rice University concluded that increased shale gas production in the U.S. and Canada could help prevent Russia and Persian Gulf countries from dictating higher prices for the gas they export to European countries.

The Obama administration believes that increased shale gas development will help reduce greenhouse gas emissions (in 2012, U.S. carbon dioxide emissions dropped to a 20-year low). Human and public health will both benefit from shale gas displacing coal burning.

A 2013 review by the United Kingdom Department of Energy and Climate Change noted that most studies of the subject have estimated that life-cycle greenhouse gas (GHG) emissions from shale gas are similar to those of conventional natural gas, and are much less than those from coal, usually about half the greenhouse gas emissions of coal; the noted exception was a 2011 study by Howarth and others of Cornell University, which concluded that shale GHG emissions were as high as those of coal.

Some 2011 studies pointed to high rates of decline of some shale gas wells as an indication that shale gas production may ultimately be much lower than is currently projected. But shale-gas discoveries are also opening up substantial new resources of tight oil / "shale oil".

North America has been the leader in developing and producing shale gas. The economic success of the Barnett Shale play in Texas in particular has spurred the search for other sources of shale gas across the United States and Canada.

A 2011 New York Times investigation of industrial emails and internal documents found that the financial benefits of unconventional shale gas extraction may be less than previously thought, due to companies intentionally overstating the productivity of their wells and the size of their reserves. The article was criticized by, among others, the New York Time's own Public Editor for lack of balance in omitting facts and viewpoints favorable to shale gas production and economics.

In first quarter 2012, the United States imported 840 billion cubic feet (Bcf) (785 from Canada) while exporting 400 Bcf (mostly to Canada); both mainly by pipeline. Almost none is exported by ship as LNG, as that would require expensive facilities. In 2012, prices went down to $3/MMBtu due to shale gas.

A recent academic paper on the economic impacts of shale gas development in the US finds that natural gas prices have dropped dramatically in places with shale deposits with active exploration. Natural gas for industrial use has become cheaper by around 30% compared to the rest of the U.S. This stimulates local energy intensive manufacturing growth, but brings the lack of adequate pipeline capacity in the U.S. in sharp relief.

One of the byproducts of shale gas exploration is the opening up of deep underground shale deposits to "tight oil" or shale oil production. By 2035, shale oil production could "boost the world economy by up to $2.7 trillion, a PricewaterhouseCoopers (PwC) report says. It has the potential to reach up to 12 percent of the world's total oil production — touching 14 million barrels a day — 'revolutionizing' the global energy markets over the next few decades."

Words and Expressions

shale gas	页岩气
Energy Information Administration	能源信息管理局
Persian Gulf	波斯湾
greenhouse gas	（GHG）温室气体
conventional [kən'venʃənl]	*adj.* 传统的，平常的，依照惯例的，约定的
Cornell University	美国康奈尔大学
tight oil	致密油
spur [spɜ:]	*vt.* 加速，鞭策；*vi.* 策马飞奔，急速前进
MMBtu=million British Thermal Units	代表百万英热单位，百万英制热单位
Pricewaterhouse Coopers	普华永道，四大国际会计师事务所之一

Lesson Two   Catalytic Cracking
催化裂化

How does fracking work?

Catalytic cracking breaks complex hydrocarbons into simpler molecules in order to increase the quality and quantity of lighter, more desirable products and decrease the amount of residuals. This process rearranges the molecular structure of hydrocarbon compounds to convert heavy hydrocarbon feedstock into lighter fractions such as kerosene, gasoline, liquified petroleum gas (LPG), heating oil, and petrochemical feedstock.

Catalytic cracking is similar to thermal cracking except that catalysts facilitate the conversion of the heavier molecules into lighter products. Use of a catalyst (a material that assists a chemical reaction but does not take part in it) in the cracking reaction increases the yield of improved-quality products under much less severe operating conditions than in thermal cracking. Typical temperatures are from 850-950°F at much lower pressures of 10-20 psi. The catalysts used in refinery cracking units are typically solid materials (zeolite, aluminum hydrosilicate, treated bentonite clay, fuller's earth, bauxite, and silica-alumina) that come in the form of powders, beads, pellets or shaped materials.

There are three basic functions in the catalytic cracking process:

Reaction: Feedstock reacts with catalyst and cracks into different hydrocarbons;

Regeneration: Catalyst is reactivated by burning off coke;

Fractionation: Cracked hydrocarbon stream is separated into various products.

The three types of catalytic cracking processes are fluid catalytic cracking (FCC), moving-bed catalytic cracking, and Thermofor catalytic cracking (TCC). The catalytic cracking process is very flexible, and operating parameters can be adjusted to meet changing product demand. In addition to cracking, catalytic activities include dehydrogenation, hydrogenation, and isomerization.

Fluid catalytic cracking or "cat cracking," is the basic gasoline-making process. Using intense heat (about 1,000 degrees Fahrenheit), low pressure and a powdered catalyst (a substance that accelerates chemical reactions), the cat cracker can convert most relatively heavy fractions into smaller gasoline molecules. The fluid cracker consists of a catalyst section and a fractionating section that operate together as an integrated processing unit (Fig. 2-2). The catalyst section contains the reactor and regenerator, which, with the standpipe and riser, forms the catalyst circulation unit. The fluid catalyst is continuously circulated between the reactor and the regenerator using air, oil vapors, and steam as the conveying media.

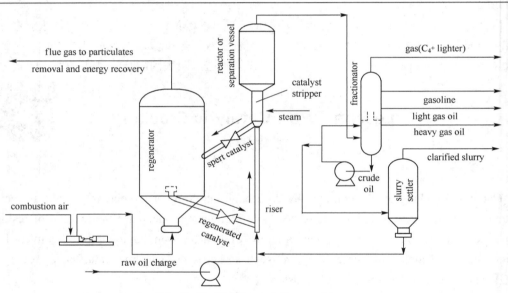

Fig.2-2 Fluid catalytic cracking

A typical FCC process involves mixing a preheated hydrocarbon charge with hot, regenerated catalyst as it enters the riser leading to the reactor. The charge is combined with a recycle stream within the riser, vaporized, and raised to reactor temperature (900-1,000° F) by the hot catalyst. As the mixture travels up the riser, the charge is cracked at 10-30 psi. In the more modern FCC units, all cracking takes place in the riser. The "reactor" no longer functions as a reactor; it merely serves as a holding vessel for the cyclones. This cracking continues until the oil vapors are separated from the catalyst in the reactor cyclones. The resultant product stream (cracked product) is then charged to a fractionating column where it is separated into fractions, and some of the heavy oil is recycled to the riser.

Spent catalyst is regenerated to get rid of coke that collects on the catalyst during the process. Spent catalyst flows through the catalyst stripper to the regenerator, where most of the coke deposits burn off at the bottom where preheated air and spent catalyst are mixed. Fresh catalyst is added and worn-out catalyst removed to optimize the cracking process.

UOP has leveraged its FCC experience and know-how to develop and license a new type of cracking process. The PetroFCC^{TM} process targets the production of petrochemical feedstocks rather than fuel products. The new process, which utilizes a uniquely designed FCC unit, can produce very high yields of light olefins and aromatics when coupled with an aromatics complex (shown in Fig. 2-3).

Driven by an increased demand for polyethylene and poly-propylene, future demand is expected to increase for petrochemical feedstocks—particularly the light olefins, ethylene and propylene. Ethylene demand is projected to grow by a compound rate of about 5% per year for the foreseeable future, while growth in propylene demand is expected to be even greater. The additional propylene produced from the increase in steam cracker ethylene production is expected to be insufficient to meet the demand, and propylene from other sources will be required.

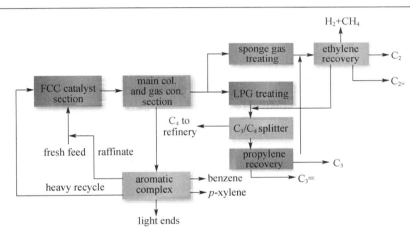

Fig.2-3   UOP petro FCC complex

Because the FCC process has proven to be a very flexible process, it is in a unique position to fill this void in supplying the expected increase in propylene demand. Although its principal function has been to produce gasoline, the FCC unit is frequently operated to maximize other products, such as distillates or LPG. The LPG mode can be considered a step towards a petrochemical mode of operation since it provides enhanced yields of the petrochemical feedstocks. However, to be considered a major component in the petrochemical complex, an FCC style unit must produce substantially greater quantities of these light olefins, produce other petrochemical feedstocks of interest, and minimize or eliminate the yield of gasoline and heavier liquid fuels. This is the targeted area for the PetroFCC process.

Words and Expressions

catalytic cracking   催化裂化
facilitate [fəˈsɪlɪteɪt]   vt. 促进，帮助

zeolite [ˈziːəlaɪt]	n.	沸石
hydrosilicate [haɪdrəˈsɪləkɪt]	n.	含水硅酸盐
bentonite [ˈbentəˌnaɪt]	n.	膨润土
fuller's earth	n.	漂白土
bauxite [ˈbɔːksaɪt]	n.	铝土岩
silica [ˈsɪlɪkə]	n.	硅土，二氧化硅
alumina [əˈluːmənə]	n.	矾土，氧化铝
fractionation [ˌfrækʃənˈeɪʃən]	n.	分馏法
parameter [pəˈræmɪtə]	n.	参数，参量，系数
isomerization [aɪˌsɒmərəˈzeɪʃən]	n.	异构化，异构化作用
Fahrenheit [ˈfɑːrənhaɪt]	n.	华氏温度计，华氏温标
cyclone [ˈsaɪkləʊn]	n.	旋风
leveraged [ˈliːvərɪdʒd]	n.	杠杆作用，影响力
olefin [ˈəʊləfɪn]	n.	[化]烯烃
aromatics [ˌærəʊˈmætɪks]	n.	芳（香）族化合物（=aromatic compound），芳香烃
raffinate [ˈræfəˌneɪt]	n.	萃余液，残油液，剩余液

Notes

1. The charge is combined with a recycle stream within the riser, vaporized, and raised to reactor temperature (900-1,000° F) by the hot catalyst.

句子分析：句子的结构为"主语+被动语态"，and 连接一个并列的句子。be combined with 的意思是与……结合。

译文：控制在提升管中的再循环物料的混合，通过热催化剂使物料蒸发，以提升反应器温度到 900~1000 华氏度。

2. The resultant product stream (cracked product) is then charged to a fractionating column where it is separated into fractions, and some of the heavy oil is recycled to the riser.

句子分析：句子的结构为"主语+被动语态"，这是一个被动语态的句子。be separated into 的意思是被分成；where 引导地点状语从句；fractions 的意思是小部分。

译文：生成物（裂化产物）经精馏塔得到不同的馏分，一些重组分油再循环至提升管（继续反应）。

3. The additional propylene produced from the increase in steam cracker ethylene production is expected to be insufficient to meet the demand, and propylene from other sources will be required.

句子分析：句子的结构为"主语+被动语态"，这是一个被动语态的句子。insufficient to 的意思是不足以；be expected to 表示有望做某事，被期待做某事。

译文：在增产乙烯时副产的丙烯无法满足增长的需求，因此需要从其他途径获得丙烯。

4. The LPG mode can be considered a step towards a petrochemical mode of operation since it provides enhanced yields of the petrochemical feedstocks.

句子分析：句子的结构为"主语+被动语态"，since 引导原因状语从句。Enhanced 的意思是增强的。

译文：液化石油气路线被认为是制备石化产品的途径之一，因为这种模式提高了石化原料的收率。

5. However, to be considered a major component in the petrochemical complex, an FCC style unit must produce substantially greater quantities of these light olefins, produce other petrochemical feedstocks of interest, and minimize or eliminate the yield of gasoline and heavier liquid fuels.

句子分析：句子的结构为"主语+谓语+宾语"，and 连接并列的谓语结构，to 引导目的状语从句。minimize or eliminate 表示减少或消除 substantially 表示大体上，充分上。

译文：然而，考虑到 FCC 装置是石油化工联合工厂重要的组成，催化裂化工艺应减小或不生产汽油和重质液体燃料，用以生产出更多的低碳烯烃和利润更高的石化原料。

Reading Comprehension

1. What are the three types of catalytic cracking processes?
2. Can you describe the process of fluid catalytic cracking?
3. What are the materials and products of FCC?
4. What is the major improvement in FCC technology?
5. What are the advantages of the PetroFCC process?

Reading Material

Semi-regenerative Catalytic Reformer　半再生催化重整

The most commonly used type of catalytic reforming unit has three reactors, each with a fixed bed of catalyst, and all of the catalyst is regenerated in situ during routine catalyst regeneration shutdowns which occur approximately once each 6 to 24 months. Such a unit is referred to as a semi-regenerative catalytic reformer (SRR).

The process flow diagram（Fig.2-4） below depicts a typical semi-regenerative catalytic reforming unit.

The liquid feed (at the bottom left in the diagram) is pumped up to the reaction pressure (5-45 atm) and is joined by a stream of hydrogen-rich recycle gas. The resulting liquid–gas mixture is preheated by flowing through a heat exchanger. The preheated feed mixture is then totally vaporized and heated to the reaction temperature (495–525°C) before the vaporized reactants enter the first reactor. As the vaporized reactants flow through the fixed bed of catalyst in the reactor, the major reaction is the dehydrogenation of naphthenes to aromatics which is highly endothermic and results in a large temperature decrease between the inlet and outlet of the reactor. To maintain the required reaction temperature and the rate of reaction, the vaporized stream is reheated in the second fired heater before it flows through the second reactor. The temperature again decreases across the second reactor and the vaporized stream must again be reheated in the third fired heater before it flows through the third reactor. As the vaporized stream proceeds through the three reactors, the reaction rates decrease and the reactors therefore become larger. At the same time, the amount of reheat required between the reactors becomes smaller. Usually, three reactors are all that is required to provide the desired performance of the catalytic reforming unit.

Some installations use three separate fired heaters as shown in the schematic diagram and some installations use a single fired heater with three separate heating coils.

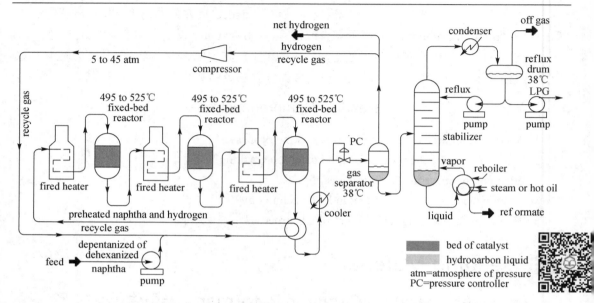

Fig.2-4 A typical semi-regenerative catalytic reforming unit

The hot reaction products from the third reactor are partially cooled by flowing through the heat exchanger where the feed to the first reactor is preheated and then flow through a water-cooled heat exchanger before flowing through the pressure controller (PC) into the gas separator.

Most of the hydrogen-rich gas from the gas separator vessel returns to the section of the recycle hydrogen gas compressor and the net production of hydrogen-rich gas from the reforming reactions is exported for use in the other refinery processes that consume hydrogen (such as hydrodesulfurization units and/or a hydrocracker unit).

The liquid from the gas separator vessel is routed into a fractionating column commonly called a stabilizer. The overhead offgas product from the stabilizer contains the byproduct methane, ethane, propane and butane gases produced by the hydrocracking reactions, and it may also contain some small amount of hydrogen. That offgas is routed to the refinery's central gas processing plant for removal and recovery of propane and butane. The residual gas after such processing becomes part of the refinery's fuel gas system.

The bottoms product from the stabilizer is the high-octane liquid reformate that will become a component of the refinery's product gasoline. Reformate can be blended directly in the gasoline pool but often it is separated in two or more streams. A common refining scheme consists in fractionating the reformate in two streams, light and heavy reformate. The light reformate has lower octane and can be used as isomerization feedstock if this unit is available. The heavy reformate is high in octane and low in benzene, hence it is an excellent blending component for the gasoline pool.

Benzene is often removed with a specific operation to reduce the content of benzene in the reformate as the finished gasoline has often an upper limit of benzene content (in the UE this is 1% volume). The benzene extracted can be marketed as feedstock for the chemical industry.

Words and Expressions

in situ	在原位置，在原处
shutdown ['ʃʌtdaʊn]	n. 关闭，倒闭，关机，停工
depict [dɪ'pɪkt]	vt. 描绘，描画，描述
dehydrogenation [diːˌhaɪdrədʒə'neɪʃən]	n. 脱氢作用
naphthene ['næfθiːn]	n. 环烷烃
aromatic [ˌærəʊ'mætɪk]	n. 芳烃
endothermic [ˌendəʊ'θɜːmɪk]	adj. 吸热的，吸能的
suction ['sʌkʃən]	n. 吸，抽吸
compressor [kəm'presə]	n. 压气机，压缩机
net production	净产量，净生产量
hydrodesulfurization [haɪdrədiːsʌlfjʊrɪ'zeɪʃən]	n. 加氢脱硫（过程），氢化脱硫作用
hydrocracker ['haɪdrəˌkrækə]	n. 加氢裂化装置
a fractionating column	精馏塔
stabilizer ['steɪbɪlaɪzə]	n. 稳定塔
isomerization [aɪˌsɒmərə'zeɪʃne]	n. 异构化（作用）
EU= European Union	欧洲联盟（简称欧盟）

Lesson Three Thermal Cracking of Hydrocarbons
热裂解制乙烯

The thermal cracking reaction is known as homolytic fission and produces alkenes, which are the basis for the economically important production of polymers. Thermal cracking is currently used to "upgrade" very heavy fractions or to produce light fractions or distillates, burner fuel and/or petroleum coke. The extremes of the thermal cracking in terms of product range is represented by the high-temperature process called "steam cracking" or pyrolysis (ca. 750 °C to 900 °C or higher) which produces valuable ethylene and other feedstocks for the petrochemical industry.

During the evolution of the thermal cracking process, various parameters, such as the average residence time and average hydrocarbon partial pressure, were found to be important. Computing systems employing complicated mathematical techniques are used to solve stiff differential equations. As a result, operations can be optimized relative to product distribution, selectivity, yield, and economic profits. Design and operation of the thermal cracking units are of critical importance. Many thermal cracking units employ two furnaces or reaction zones with one convection zone (see Fig. 2-5). Thermal cracking occurs in the coils in the radiant zone. In the convection zones, heat is transferred as follows: first, to preheat and vaporize the hydrocarbons to be cracked; second, to preheat the steam to be mixed with the hydrocarbon feeds; third, to preheat the entering fuel (often natural gas) and air (used to burn the fuel); fourth, to preheat the boiler feed water; fifth, to produce steam from boiler feed water; and sixth, to superheat the steam. Preheating of the fuel and air results in higher flame temperatures, which increases both the rate and the amount of heat transfer to the reacting gases in the coils.

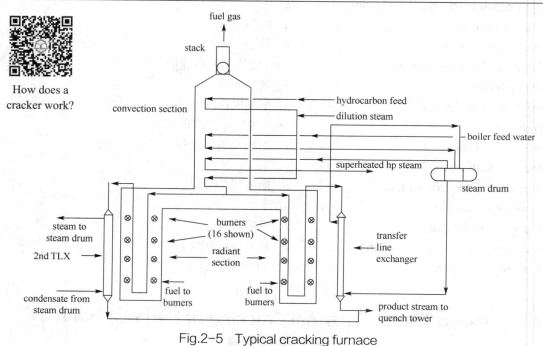

Fig.2-5 Typical cracking furnace

Product Separation, Recovery, and Purification (Fig.2-6) represents a simplified schematic flow diagram for an ethane and/or propane cracking ethylene plant. The separation sequences are shown as follows:

(1) Quench tower. This is accomplished by quenching the product effluent stream with water. Condensed hydrocarbons and water from the quench tower are separated in a coalescer. Hydrocarbons from this coalescer and the drips from the first stage compressor discharge drums are sent to the drip stripper to separate lighter hydrocarbons from the pyrolysis gasoline and fuel oil. Water from the coalescer is transferred to the process water stripper to separate the hydrocarbons before sending the water to the biopond for wastewater treatment. Quench tower design is a

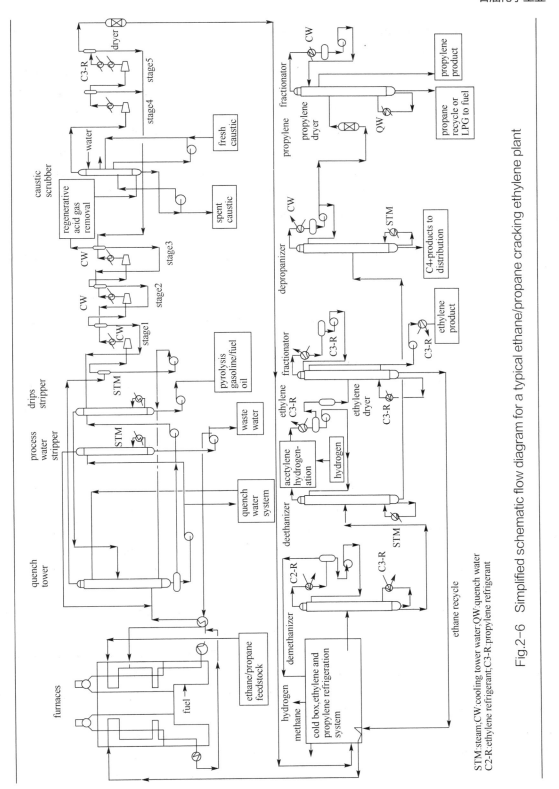

Fig.2-6 Simplified schematic flow diagram for a typical ethane/propane cracking ethylene plant

STM:steam;CW:cooling tower water;QW:quench water
C2-R:ethylene refrigerant;C3-R:propylene refrigerant

challenge. A considerable amount of water is recirculated, and coke particles in the water make this stream quite erosive requiring special pumps.

(2) Compression and condensation. The pyrolysis gas leaving the quench tower is compressed to approximately 35 atmospheres in a four or five stage centrifugal compressor (normally referred to as cracked gas compressor). The number of stages is determined by the maximum temperature for the material of construction of the cracked gas compressor and the fouling tendency of the pyrolysis gas. The compressor consists of two or three compressor casings driven by a single or double extraction/condensing turbines depending on plant size. Between each stage of the compressor are condensers and drums that separate the condensed hydrocarbons and water.

(3) Acid gas removal. Acid gases (CO_2 and H_2S) are removed by absorption after the third or the fourth compression stage. This is the optimum location since the gas volume is significantly reduced. Scrubbing with caustic solution, monoethanolamine (MEA), or diethanolamine (DEA) is generally used.

(4) Water removal. Complete removal of water vapor from the pyrolysis gas is generally achieved using dryers packed with molecular sieves. One dryer is operated while the other is in the regeneration mode.

(5) Cold box and refrigeration system. After the acid gas and water removal, the pyrolysis gas is cooled and condensed to approximately $-165℃$; only hydrogen and some methane remain in the vapor phase. The feed locations are determined via process simulation. Hydrogen and methane are drawn from the lowest temperature stage separator and sent to thermal cracking furnaces as fuel.

(6) The demethanizer is designed for complete separation of methane from ethylene and heavier components. The demethanizer is normally operated at approximately 7 atm. The demethanizer overhead consists of methane, plus relatively small amounts of hydrogen, carbon monoxide, and traces of ethylene. Brazed aluminum plate-fin exchangers are used for the multipass cryogenic heat-transfer services. Some ethylene plants employ high pressure demethanizer operated at approximately 35 atm. They are generally combined with either a front-end deethanizer or depropanizer.

(7) Deethanizer and ethylene fractionator (ethylene/ethane splitter). The C_2 and heavier hydrocarbons from the bottom of the demethanizer are sent to the deethanizer operated at approximately 25 atm. It is either a trayed tower or a packed column. Deethanizer overhead consists of C_2 hydrocarbons and the bottom products are C_3 and heaviers.

(8) Acetylene hydrogenation. In modern ethylene plants, acetylene is generally hydrogenated to ethylene and ethane in a palladium catalyst bed. The reaction is quite exothermic, and intermediate cooling is required. The effluent normally contains less than 1 ppmv of acetylene. This is generally referred to as back-end acetylene hydrogenation, which exerts higher selectivity and more precise temperature control. Front-end acetylene hydrogenation is also practiced in the stream of an intermediate compressor. Front-end acetylene hydrogenation uses a different catalyst, and a deethanizer or a depropanizer column

is ahead of the demethanizer; this scheme has become of increased importance in the recent past. After acetylene hydrogenation, the dried gas enters the ethylene fractionator to separate ethylene from ethane. A section is provided near the top of the fractionator for removal of residual hydrogen, carbon monoxide, and methane. A closed loop heat pump is sometimes used, which uses propylene refrigerant as the coolant in the reflux condenser. Ethane is normally recycled back to the thermal cracking furnaces to be cracked.

(9) Depropanizer and propylene fractionator (propylene/propane splitter). The deethanizer bottom is sent to the depropanizer where the C_3 hydrocarbons are the overhead product and the C_4 and heavier hydrocarbons are removed from the bottom. The overhead of the depropanizer is sent to the propylene fractionator where propylene is separated from propane. A two-tower propylene fractionator produces polymer grade propylene (99.5% plus). For the naphthas or gas oil cracking plants, the depropanizer bottom stream is normally further processed in a debutanizer, which separates the C_4 and lighter in the overhead, and the pyrolysis gasoline and heavier in the bottom. Also, the depropanizer overhead stream generally is sent to a methylacetylene and propadiene (MAPD) hydrogenation reactor, where MA is hydrogenated to propylene and propane.

Words and Expressions

homolytic	adj. 均裂的
fission ['fɪʃən]	n. [物]（原子的）分裂，裂变
alkene ['ælkiːn]	n. 烯烃，链烯
pyrolysis [ˌpaɪə'rɒlɪsɪs, ˌpɪə'rɒlɪsɪs]	n. 高温分解
ca.=circa(about)	大约；（时间）前后
thermal cracking	热裂解
differential equation	微分方程
convection [kən'vekʃən]	n. 对流，传送
boiler feed water	锅炉给水
superheat [ˌsjuːpə'hiːt]	vt. 使……过热；n. 过热
purification [ˌpjʊərɪfɪ'keɪʃən]	n. 净化，提纯
flow diagram	流程图
quench tower	急冷塔
effluent ['efluənt]	n. 流出物，污水，废气
coalescer [ˌkəʊə'lesə]	n. 聚结器
pyrolysis gasoline	热解汽油
stripper ['strɪpə]	n. 汽提塔
erosive [ɪ'rəʊsɪv]	adj. 腐蚀的，侵蚀性的
condensation [kɒnden'seɪʃən]	n. 压缩，凝结，冷凝
biopond [baɪəpɒnd]	n. 生物池（生物废水处理装置）
centrifugal compressor	离心式压缩机
fouling ['faʊlɪŋ]	n. 沉积物

monoethanolamine [mɒnoʊeθənoʊ'læmiːn]   n. 单乙醇胺
diethanolamine [dɪrθənoʊ'læmiːn]   n. 二乙醇胺
molecular sieves   分子筛
acetylene [ə'setɪliːn]   n. 乙炔
palladium [pə'leɪdɪəm]   n. 钯
ppmv   按体积计算百万分之一
reflux ['riːflʌks]   n. 回流

Notes

1. Hydrocarbons from this coalescer and the drips from the first stage compressor discharge drums are sent to the drip stripper to separate lighter hydrocarbons from the pyrolysis gasoline and fuel oil.

译文：聚结器来的烃类物质和第一压缩段的出口罐来的凝液一起送入凝液汽提塔，可从热解汽油和燃料油中分离出轻烃。

2. The compressor consists of two or three compressor casings driven by a single or double extraction/condensing turbines depending on plant size.

driven by 的意思是由……驱动；depend on 意为依靠，依赖，取决于。

译文：由一个或两个抽出/冷凝式汽轮机（根据工厂规模）驱动的两个或三个压缩机组成一个压缩段。

3. Between each stage of the compressor are condensers and drums that separate the condensed hydrocarbons and water.

译文：在每个压缩段之间有冷凝器和出口罐用来分离冷凝下来的烃类和水。

4. After acetylene hydrogenation, the dried gas enters the ethylene fractionator to separate ethylene from ethane.

译文：在加氢脱炔装置后，干燥气进入乙烯精馏塔，分离乙烯、乙烷。

5. A section is provided near the top of the fractionator for removal of residual hydrogen, carbon monoxide, and methane.

译文：在靠近（乙烯）精馏塔的顶部有一段是用来去除残余的氢、一氧化碳和甲烷的。

Reading Comprehension

1. What's the characteristic of the thermal cracking process?
2. Please draw a diagram to explain how the Typical cracking furnace works.
3. Can you describe the product separation, recovery, and purification for an ethane and/or propane cracking ethylene plant?
4. Can you get the meaning of front-end deethanizer process and front-end depropanizer process from the context?
5. Can you find the Quench tower /Compression and condensation/ Acid gas removal/ Water removal/ Cold box and refrigeration system/ Deethanizer and ethylene fractionator demethanizer/ Acetylene hydrogenation/ Depropanizer on the following diagram (Fig. 2-6)?

Reading Material

The Isomar™ Process 异构化工艺过程

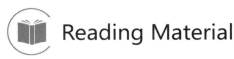

Fig.2-7 Process chemistry

The process chemistry of the isomerization reactions is shown in Fig.2-7. An Isomar unit is always combined with a recovery unit for one or more xylene isomers. Usually it is combined with a Parex unit for recovery of para-xylene. In the Parex-Isomar flow scheme, fresh mixed xylenes are fed to the xylene column, which can be designed either to recover ortho-xylene in the bottoms or simply reject C_{9+} aromatic components in order to meet feed specifications to the Parex unit. The xylene column overhead is then directed to the Parex unit, where 99.9 wt% paraxylene is produced at 97 wt% recovery per pass. The raffinate from the Parex unit, containing less than 1 wt% para xylene, is sent to the Isomar unit.

The general flow scheme of the Isomar unit is shown in the following diagram (Fig. 2-8). The feed to an Isomar unit is first combined with hydrogen-rich recycle gas and makeup gas to replace the small amount of hydrogen consumed in the Isomar reactor. The combined feed is then preheated and vaporized by exchange with reactor effluent, and raised to reactor operating temperature in a fired heater. The hot feed vapor is sent to the reactor, where it is passed through the catalyst. The reactor effluent is cooled by exchange with the combined feed and is then sent to the product separator. Hydrogen-rich gas is taken off the top of the product separator and recycled back to the reactor. A small portion of the recycle gas can be purged to remove accumulated light ends from the recycle gas loop. Liquid from the bottom of the product separator is charged to the deheptanizer column. The C_{7-} overhead from the deheptanizer is cooled and separated into gas and liquid products. The deheptanizer overhead gas is typically used as fuel in an integrated aromatics complex, and the overhead liquid is recycled back for recovery of benzene. The C_{8+} fraction from the bottom of the deheptanizer is, if necessary, combined with fresh mixed xylenes feed and recycled back to the xylene column.

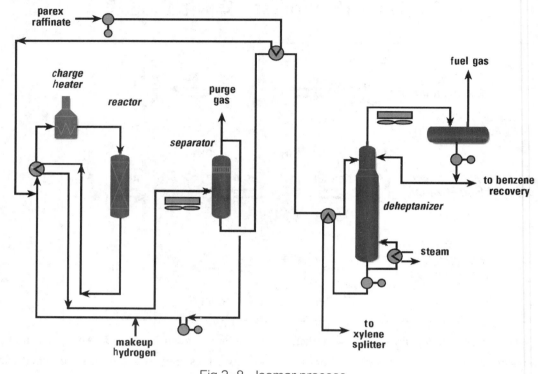

Fig.2-8 Isomar process

Words and Expressions

Isomar	UOP 公司注册商标
xylene ['zaɪliːn]	n. [化]二甲苯
dealkylation [diːælkɪ'leɪʃən]	n. 脱烷基化作用
parex	吸附分离（同 adsorptive separation）
per pass	单程
makeup ['mek,ʌp]	n. 组成，补充
reactor effluent	反应器流出物
light ends	轻馏分
deheptanizer column	脱庚烷塔

 # History, Inheritance and Development

China's Oil Refining Industry: From Quantity to Quality

After a period of rapid development, China's refining industry is moving into a high-quality development stage towards a healthier industrial structure, supported and guided by favorable government policies. Total capacity is expected to reach ca. 1 billion tons by 2025.

Market Dynamics

Starting from 1978, China's oil refining industry experienced four major development stages: After an early stage, the scale of SOE's refining capacity expanded rapidly in the 2020s due to rising demand and extensive new construction (esp. Sinopec and PetroChina). Afterwards, the industry developed steadily, and so-called Teapot refineries entered the picture as China opened the crude oil import quota to private refineries in 2015. Nowadays, the capacity of private refineries accounts for around 30% of overall 920 million tons. Recently, the government has been starting to limit capacity expansion due to environmental and profitability concerns, but still targets to reach a capacity of 1 billion tons by 2025.

Four key trends are shaping the industry in the coming years:

(1) Capacity integration: New refining projects with a capacity less than 10 million tons will be strictly limited, and small teapots will either be gradually combined or eliminated.

(2) Consumption shift: China's oil demand is expected to peak in 2030 and then to gradually decrease. Thus, the oil consumption focus gradually shifts from fuels to feedstock for petrochemical production (e.g. Naphtha).

(3) Refining-chemical integration: Flat demand for transportation fuels and rising need for petrochemical raw materials also leads to integrated refinery developments being encouraged by the government – oriented towards demand, profitability and sustainability (China targets to be carbon neutral by 2060).

(4) Import decrease: China's crude oil demand mainly relies on import and its dependency continued to increase until 2021, when crude oil imports decreased for the first time in the past 20 years.

Some of these trends are encouraged and supported by government policies in China's various regional and industry-specific 14th Five-Year Plans as well as in the Action Plan for Carbon

Dioxide Peaking Before 2030.

(1) Limited capacity expansion: "Until 2025, domestic primary processing capacity of crude oil should not exceed 1 billion tons and capacity utilization rate for main oil products should be >80%."

(2) Stricter control on pollution: "Improve the ability of utilizing and disposing hazardous wastes such as spent catalyst."

(3) More support on pilot refineries: "Develop an array of pilot enterprises with ecological guiding force and core competence in petrochemical, chemical and other industries."

(4) "Refining-chemical" integration: "Effectively promote 'fuel decreasing & chemical increasing' and extend the industrial chain of the petrochemical industry."

Already being one of the major oil refining countries worldwide, these developments will further alter China's industrial structure and domestic petrochemical capabilities on its way towards self-sufficiency and a greener future. But capacity upgrading and integration will not always be easy: Refining clusters in Shandong, Liaoning and Zhejiang will feel the effects the most.

Practice and Training

ASTM D86

Distillation of Petroleum Products at Atmospheric Pressure – Significance, Instrumentation and Standard Test Method D86

Atmospheric Distillation is one of the most important of all fundamental physical chemical properties of petroleum and petroleum products. ASTM D86 Standard Test Method is one of the oldest test methods in ASTM D2 Committee of Petroleum Products and Lubricants. It relates to the volumetric composition, energy content and boiling range distribution of fuels and petroleum products in a sense that it relates not only to volatility but also to the fuel performance the higher the boiling point components the higher the tendency to form combustion deposits giving rise to obstructions of fuel lines and failure whereas a disproportion of light components will produce irregularities in the mixture during warm up and general performance increasing the tendency to vapor locks especially at high temperatures and high altitudes.

During storage high volatile components in the fuel are particularly responsible for the formation of potentially explosive vapors. So in order to better specify the volatility characteristics of good quality products it is necessary to specify the distillation yield at several temperatures particularly at the initial boiling point at T10 (Temperature at 10% recovery), T50 (Temperature at 50% volume distilled), T90 (Temperature at 90% volume distilled) and final boiling point. These distillation values are including as Distillation Limits" in petroleum product specifications in commercial purchasing commercial agreements in refinery process quality control and to demonstrate compliance to environmental and federal regulations.

Another common way to express the volatility of a hydrocarbon product is the evaporation loss at a specified temperature for instance E200 is what percentage in volume is evaporated at 200 °C.

It is important to keep in mind that D86 is a standard method designed for the analysis of distillates, therefore is not applicable to materials containing appreciable quantities of residual fractions. D86 is used alone or in combination with other properties such as density, sulfur content to fully characterize refining products in terms of quality, quantity and composition. In the industry it is compulsory to make sure there is no adulterations and contamination that might affect the required characteristics of the final products.

Instrumentation

The determination of the Distillation Characteristics" of liquid hydrocarbon mixtures is generally performed in distillation apparatus consisting mainly of a distillation flask that holds the sample to be tested, a condenser immersed in a cooling bath to condense the vapors produced at the current temperature, a heat source, a receiving graduated cylinder and a calibrated temperature meter device. A typical arrangement can be seen in Fig. 2-9 and a commercial version shown in Fig.2-10.

Distillation lab

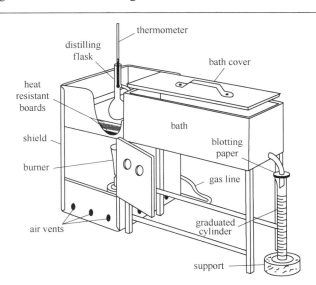

Fig. 2-9　Typical arrangement of a manual unit for atmospheric distillation

Technical Specifications Technical of Manual Laboratory Single-Batch Unit

Consisting of: an electric heating for temperature up 400℃ (distillation group 0-4), mounting on left side, made of stainless steel with inspection tempered window can be opened, electronic regulator, adjustable flask support with external control knob. Stainless steel cooling bath with nickel-plated condenser tube inclined to the right, glossy-black PVC cover, connections allowing water circulation, drain valve.

Power: 1000 W

Power supply: 230(1±10%)　50Hz

Temperature: from ambient to 400℃

Dimensions: 39cm×65cm×60cm

Weight: 9 kg

conform to ASTM D86 D216 D447 D850 D1078 E133 IP 123, ISO 3405,DIN 51751

Figure 2-10　Manual laboratory single-batch unit

Unit Three

Polymer Chemistry

聚合物化学

Lesson One Basic Concepts of Polymers
聚合物的基本概念

A polymer is a large molecule built up by the repetition of small, simple chemical units. In some cases the repetition is linear, much as a chain is built up from its links. In other cases the chains are branched or interconnected to form three-dimensional networks. The repeat unit of the polymer is usually equivalent or nearly equivalent to the monomer, or starting material from which the polymer is formed. Thus the repeat unit of poly (vinyl chloride) is —CH_2CHCl—; Its monomer is vinyl chloride, CH_2=$CHCl$.

The length of the polymer chain is specified by the number of repeat units in the chain. This is called the degree of polymerization. The molecular weight of the polymer is the product of the molecular weight of the repeat unit and the degree of polymerization, DP. Using poly (vinyl chloride) as an example, a polymer of DP. 1000 has a molecular weight of $63 \times 1000 = 63000$. Most high polymers useful for plastics, rubbers, or fibers have molecular weights between 10,000 and 1,000,000.

Unlike many products whose structure and reactions were well known before their industrial application, some polymers were produced on an industrial scale long before their chemistry or physics was studied. Empiricism in recipes, processes, and control tests was usual.

Gradually the study of polymer properties began. Almost all were first called anomalous because they were so different from the properties of low-molecular-weight compounds. It was soon realized, however, that polymer molecules are many times larger than those of ordinary substances. The presumably anomalous properties of polymers were shown to be normal for such materials, as the consequences of their size were included in the theoretical treatments of their properties.

Primary chemical bonds along polymer chains are entirely satisfied. The only forces between molecules are secondary bond forces of attraction, which are weak relative to primary bond forces. The high molecular weight of polymers allows these forces to build up enough to impart excellent strength, dimensional stability, and other mechanical properties to the substances.

There are a number of methods of classifying polymers. One is to adopt the approach of using their response to thermal treatment and to divide them into thermoplastics and thermosets. Thermoplastics are polymers which melt when heated and resolidify when cooled, while thermosets are those which do not melt when heated but, at sufficiently high temperatures, decompose irreversibly. This system has the benefit that there is a useful chemical distinction between the two groups. Thermoplastics comprise essentially linear or lightly branched polymer molecules, while thermosets are substantially crosslinked materials, consisting of an extensive three-dimensional network of covalent chemical bonding.

Words and Expressions

molecule [ˈmɒlɪkjuːl, ˈməʊlɪkjuːl]	n. 分子
repetition [ˌrepɪˈtɪʃən]	n. 重复
polymer [ˈpɒlɪmə]	n. 聚合物
linear [ˈlɪnɪə]	adj. 线的，线性的
branched [brɑːntʃt]	adj. 支化的
three-dimensional network	三维网状结构
macromolecule [ˌmækrəʊˈmɒlɪkjuːl]	n. 高分子
the degree of polymerization	聚合度
molecular-weight	n. 分子量
monomer [ˈmɒnəmə]	n. 单体
application [ˌæplɪˈkeɪʃən]	n. 应用
empiricism [emˈpɪrɪsɪzəm]	n. 经验主义
process [prəˈses]	n. 过程；加工；操作
anomalous [əˈnɒmələs]	adj. 异常的；不规则的
property [ˈprɒpətɪ]	n. 性能
thermal [ˈθɜːməl]	adj. 热的
thermoplastic [ˌθɜːməˈplæstɪk]	n. 热塑性塑料
thermoset [ˈθɜːməset]	n. 热固性塑料
covalent [kəʊˈveɪlənt]	adj. 共价的

Notes

1. In some cases the repetition is linear, much as a chain is built up from its links. In other cases the chains are branched or interconnected to form three-dimensional networks.

译文：这些重复单元有的形成线型的，很像一条链；有的形成支化，有的相互连接形成三维网状结构。

2. The molecular weight of the polymer is the product of the molecular weight of the repeat unit and the degree of polymerization.

译文：聚合物的分子量是重复单元的分子量与聚合度的乘积。

3. The only forces between molecules are secondary bond forces of attraction, which are weak

relative to primary bond forces.

译文：分子间唯一能有的力是次价键的引力，这种引力比主价键的力要小。

4. The high molecular weight of polymers allows these forces to build up enough to impart excellent strength, dimensional stability, and other mechanical properties to the substances.

译文：由于聚合物的分子量大，这些力可以累加得很大，使这些物质具有非常好的强度、尺寸稳定性和其他力学性能。

Reading Comprehension

1. What is the degree of polymerization?
2. How much is its molecular weight if a polyethylene has the degree of polymerization 10000?

Reading Material

The History of the Concept of the Macromolecule
高分子概念的历史

Modern books about polymer chemistry explain that the word polymer is derived from the Greek words "poly" meaning many and "meros" meaning part. They often then infer that it follows that this term applies to giant molecules built up of large numbers of interconnected monomer units. In fact this is misleading since historically the word polymer was coined for other reasons. The concept of polymerism was originally applied to the situation in which molecules had identical empirical formulae but very different chemical and physical properties. For example, benzene (C_6H_6; empirical formula CH) was considered to be a polymer of acetylene (C_2H_2; empirical formula also CH). Thus the word "polymer" is to be found in textbooks of organic chemistry published up to about 1920 but not with its modern meaning.

Understanding of the fundamental nature of those materials now called polymers had to wait until the 1920s, when Herman Staudinger coined the word "macromolecule" and thus clarified thinking. There was no ambiguity about this new term – it meant "large molecule", again from the Greek, and these days is used almost interchangeably with the word polymer. Strictly speaking, though, the words are not synonymous. There is no reason in principle for a macromolecule to be composed of repeating structural units; in practice, however, they usually are. Staudinger's concept of macromolecules was not at all well received at first. His wife once recalled that he had "encountered opposition in all his lecture." Typical of this opposition was that of one distinguished organic chemist who declared that it was as if zoologists "were told that somewhere in Africa an elephant was found who was 1500 feet long and 300 feet high".

Staudinger's original paper opposing the prevalent colloidal view of certain organic materials was published in 1920 and contained mainly negative evidence. Firstly, he showed that the organic substances retained their colloidal nature in all solvents in which they dissolve; by contrast, inorganic colloids lose their colloidal character when the solvent is changed. Secondly, contrary to what would have been expected, colloidal character was able to survive chemical modification of the original substance.

By about 1930 Staudinger and others had accumulated much evidence in favour of the macromolecular hypothesis. The final part in establishing the concept was carried out by Wallace Carothers of the Du Pont company in the USA. He began his work in 1929 and stated at the outset that the aim was to prepare polymers of definite structure through the use of established organic reactions. Though his personal life was tragic, Carothers was an excellent chemist who succeeded brilliantly in his aim. By the end of his work he had not only demonstrated the relationship between structure and properties for a number of polymers, but he had invented materials of tremendous commercial importance, including neoprene rubber and the nylons.

Words and Expressions

coin [kɔɪn]   v. 杜撰
polymerism [pɒˈlɪmərɪzm]   n. 聚合现象
formulae [ˈfɔːmjəliː]   formula 的复数；公式，规则
acetylene [əˈsetɪliːn]   n. 乙炔
ambiguity [ˌæmbɪˈɡjuːɪtɪ]   n. 含糊；不明确
synonymous [sɪˈnɒnɪməs]   adj. 同义词的
prevalent [ˈprevələnt]   adj. 流行的；普遍的
colloidal [ˈkɒlɔɪdəl]   adj. 胶体的
solvent [ˈsɒlvənt]   n. 溶剂
neoprene [ˈniːəupriːn]   n. 氯丁橡胶
nylon [ˈnaɪlən]   n. 尼龙

Lesson Two   Polymer Structure and Physical Properties
聚合物的结构及其物理性能

Since the acceptance of the macromolecular hypothesis in the 1920's, it has been recognized that the unique properties of polymers—for example, the elasticity and abrasion resistance of rubbers, the strength and toughness of fibers, and the flexibility and clarity of films—must be attributed to their long-chain structure. In the examination of structure-property relationships it is advantageous to classify properties into those involving large and small deformations. The former class includes such properties as tensile strength and phenomena observed in the melt, while properties involving only small deformations include electrical and optical behavior, such mechanical properties as stiffness and yield point, and the glass and crystalline melting transitions.

Properties involving large deformations depend primarily on the long-chain nature of polymers and the gross configuration of their chains. Important factors for this group of properties include molecular weight and its distribution, chain branching and the related category of side-chain substitution, and crosslinking.

Physical properties associated with small deformations are influenced most by factors determining the manner in which chain atoms interact at small distances. The ability of polymers to crystallize set by considerations of symmetry and steric effects has major importance here, as do the flexibility of the chain bonds and the number, nature, and spacing of polar groups. To the extent that they influence the achievement of local order, gross configurational properties are also important. Similar considerations apply to amorphous polymers below the glass transition.

In crystalline polymers, the nature of the crystalline state introduces another set of the crystal structure, degree of crystallinity, size and number of spherulites, and orientation. Some of these phenomena are in turn influenced by the conditions of fabrication of the polymer.

Finally, the properties of polymers can be varied importantly by the addition of other materials, such as plasticizers or reinforcing fillers. Properties involving both large and small deformations may be influenced in this way.

One of the most important determinants of polymer properties is the location in temperature of the major transitions, the glass transition T_g and the crystalline melting point T_m. With few exceptions, polymer structure affects the glass transition T_g and the crystalline melting point T_m similarly. This is not unexpected, since similar considerations of cohesive energy and molecular packing apply to the amorphous and crystalline or paracrystalline regions, respectively, in accounting for the temperature levels at which the transitions occur. In consequence, T_m and T_g are rather simply related for many polymers. Depending on symmetry, T_g is approximately one-half to two-thirds T_m. There are, however, some exceptions to this general rules.

The variables necessary to define the mechanical and physical propertied of polymers have now been discussed. The increase of T_m with molecular weight, leveling off as polymer molecular weight are reached, the related approximate behavior of T_g, and the continual increase of viscosity with molecular weight serve to define, in terms of the variables of molecular weight and temperature, regions in which the properties of typical plastics, rubbers, viscous liquids, and so on, may be found.

Words and Expressions

hypothesis [haɪˈpɒθɪsɪs]	n. 学说，假说
elasticity [ɪlæsˈtɪsɪtɪ]	n. 弹性
abrasion resistance	耐磨性
toughness [ˈtʌfnɪs]	n. 韧性
flexibility [ˌfleksəˈbɪlɪtɪ]	n. 柔软性
clarity [ˈklærɪtɪ]	n. 透明性
deformation [ˌdiːfɔːˈmeɪʃən]	n. 变形
tensile strength	拉伸强度
phenomena [fɪˈnɒmɪnə]	n. 现象
optical [ˈɒptɪkəl]	adj. 光学的

mechanical properties	力学性能，机械性能
stiffness ['stɪfnɪs]	n. 刚性
yield point	屈服点
crystalline ['krɪstəlaɪn]	n. 结晶
configuration [kən,fɪgjʊ'reɪʃən]	n. 构型，构造，结构
category ['kætɪgərɪ]	n. 种类
substitution [,sʌbstɪ'tjuːʃən]	n. 取代，代替
crosslinking [k'rɒslɪŋkɪŋ]	n. 交联
amorphous [ə'mɔːfəs]	adj. 无定形的，非晶的
spherulites [sfe'ruːlaɪts]	n. 球晶
orientation [,ɔːrɪen'teɪʃən]	n. 取向
plasticizer ['plæstə,saɪzə]	n. 增塑剂
cohesive energy	内聚能
molecular packing	分子排列
paracrystalline [pæ'reɪkrɪstəlaɪn]	adj. 次晶的，亚晶状的
symmetry ['sɪmɪtrɪ]	n. 对称
variable ['veərɪəbl]	n. 变量，可变参数

Notes

1. The former class includes such properties as tensile strength and phenomena observed in the melt, while properties involving only small deformations include electrical and optical behavior, such mechanical properties as stiffness and yield point, and the glass and crystalline melting transitions.

译文：前者（与大形变相关的性能）包含拉伸强度和熔体中观测到的现象，而与小形变相关的性能有电和光性能，刚性和屈服点等力学性能，及玻璃化转变和结晶熔融转变。

2. To the extent that they influence the achievement of local order, gross configurational properties are also important. Similar considerations apply to amorphous polymers below the glass transition.

译文：整体构型特征是很重要的，在一定程度上影响着局部的有序性。在玻璃化转变以下，无定形聚合物的情况也相类似。

3. This is not unexpected, since similar considerations of cohesive energy and molecular packing apply to the amorphous and crystalline or paracrystalline regions, respectively, in accounting for the temperature levels at which the transitions occur.

译文：这是在意料中的，因为在转化温度（不论是玻璃化转变温度还是结晶熔融点温度）范围，无论是无定形、结晶还是先结晶区域，其内聚物和分子排列是相似的。

4. The increase of T_m with molecular weight, leveling off as polymer molecular weight are reached, the related approximate behavior of T_g, and the continual increase of viscosity with molecular weight serve to define, in terms of the variables of molecular weight and temperature, regions in which the properties of typical plastics, rubbers, viscous liquids, and so on, may be found.

译文：熔点随分子量的增加而提高，当聚合物分子量趋于平缓时，玻璃化温度的相关行为、黏度随分子量逐渐增加的现象都是很明显的。从分子量和温度这两变量来看，可以发现典型的塑料、橡胶和黏性液体等的各种特征。

Reading Comprehension

1. Which properties are related to the large deformations of polymer?
2. Which properties are related to the small deformations of polymer?
3. What is the determination of polymer properties?

Reading Material

The Glass Transition Temperaure, T_g 玻璃化转变温度

The glass transition is a phenomenon observed in linear amorphous polymers, such as poly(styrene) or poly(methyl methacrylate). It occurs at a fairly well-defined temperature when the bulk material ceases to be brittle and glassy in character and becomes less rigid and more rubbery.

Many physical properties change profoundly at the glass transition temperature, including coefficient of thermal expansion, heat capacity, refractive index, mechanical damping, and electrical properties. All of these are dependent on the relative degree of freedom for molecular motion within a given polymeric material and each can be used to monitor the point at which the glass transition occurs. Unfortunately, in certain cases, the values obtained from these various techniques can vary widely. An example is the variation found in the measured values of T_g for poly(methyl methacrylate), which range from 110℃ using dilatometry to 160℃ using a rebound elasticity technique. This, though, is an extreme example; despite the fact that the measured value of T_g does vary according to the technique used to evaluate it, the variation tends to be over a fairly small temperature range.

The glass transition is a second-order transition. In this it differs from genuine phase changes that substances may undergo, such as melting or boiling, which are known as first-order transitions. These latter transitions are characterized by a distinct volume change, by changes in optical properties (i.e, in the X-ray diffraction pattern and the infrared spectrum) and by the existence of a latent heat for the phase change in question. By contrast, no such changes occur at the glass transition, though the rate of change of volume with temperature alters at the T_g, as illustrated in Fig. 3-1.

The glass transition can be understood by considering the nature of the changes that occur at the temperature in question. As a material is heated to this point and beyond, molecular rotation around single bonds suddenly becomes significantly easier. A number of factors can affect the ease with which such molecular rotation takes place, and hence influence the actual value that the glass transition temperature takes. The inherent mobility of a single polymer molecule is important and molecular features which either increase or reduce this mobility will cause differences in the value of T_g. In addition, interactions between polymer molecules can lead to restrictions in molecular mobility, thus altering the T_g of the resulting material.

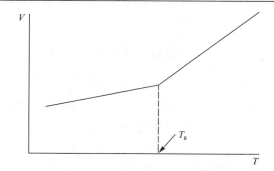

Glass transition temperature

Fig. 3-1 Plot of volume against temperature for a typical polymer passing through its glass transition

Words and Expressions

amorphous [ə'mɔːfəs]	*adj.* 无定形的
styrene ['staɪriːn]	*n.* 苯乙烯
methacrylate [me'θækrə,leɪt]	*n.* 甲基丙烯酸酯
refractive [rɪ'fræktɪv]	*adj.* 折射的
damp [dæmp]	*v.* 阻尼
dilatometry [dɪleɪ'tʊmɪtrɪ]	*n.* 膨胀法

Lesson Three Applications of Polymers
聚合物的应用

Macromolecular science has had a major impact on the way we live. It is difficult to find an aspect of our lives that is not affected by polymers. Just 50 years ago, materials we now take for granted were non-existent. With further advances in the understanding of polymers, and with new applications being researched, there is no reason to believe that the revolution will stop any time soon.

This section presents some common applications of the polymer classes introduced in the section on Polymer Structure. These are by no means all of the applications, but a cross section of the ways polymers are used in industry.

Elastomers

Rubber is the most important of all elastomers. Natural rubber is a polymer whose repeating unit is isoprene. This material, obtained from the bark of the rubber tree, has been used by humans for many centuries. It was not until 1823, however, that rubber became the valuable material we know today. In that year, Charles Goodyear succeeded in "vulcanizing" natural rubber by heating it with sulfur. In this process, sulfur chain fragments attack the polymer chains and lead to cross-linking. The term vulcanization is often used now to describe the cross-linking of all elastomers.

Much of the rubber used in the United States today is a synthetic variety called styrene-butadiene rubber (SBR). Initial attempts to produce synthetic rubber revolved around isoprene because of its presence in natural rubber. Researchers eventually found success using butadiene and styrene with sodium metal as the initiator. This rubber was called Buna-S—"Bu" from butadiene, "na" from the symbol for sodium, and "S" from styrene. During Word War II, hundreds of thousands of tons of synthetic rubber were produced in government controlled factories. After the war, private industry took over and changed the name to styrene-butadiene rubber. Today the United States consumes on the order of a million tons of SBR each year. Natural and other synthetic rubber materials are quite important.

Plastics

Americans consume approximately 60 billion pounds of plastics each year. The two main types of plastics are thermoplastics and thermosets. Thermoplastics soften on heating and harden on cooling while thermoset, on heating, flow and cross-link to form rigid material which does not soften on future heating. Thermoplastics account for the majority of commercial usage.

Among the most important and versatile of the hundreds of commercial plastics is polyethylene. Polyethylene is used in a wide variety of applications because, based on its structure, it can be produced in many different forms. The first type to be commercially exploited was called low density polyethylene (LDPE) or branched polyethylene. This polymer is characterized by a large degree of branching, forcing the molecules to be packed rather loosely forming a low density material. LDPE is soft and pliable and has applications ranging from plastic bags, containers, textiles, and electrical insulation, to coatings for packaging materials.

Another form of polyethylene differing from LDPE only in structure is high density polyethylene (HDPE) or linear polyethylene. This form demonstrates little or no branching, enabling the molecules to be tightly packed. HDPE is much more rigid than branched polyethylene and is used in applications where rigidity is important. Major uses of HDPE are plastic tubing, bottles, and bottle caps.

Other forms of this material include high and ultra-high molecular weight polyethylenes. HMW and UHMW, as they are known. These are used in applications where extremely tough and resilient materials are needed.

Fibers

Fibers represent a very important application of polymeric materials, including many examples from the categories of plastics and elastomers.

Polyester yarn manufancturing process

Natural fibers such as cotton, wool, and silk have been used by humans for many centuries. In 1885, artificial silk was patented and launched the modern fiber industry. Manmade fibers include materials such as nylon, polyester, rayon, and acrylic. The combination of strength, weight, and durability has made these materials very important in modem industry.

Generally speaking, fibers are at least 100 times longer than they are wide. Typical natural and artificial fibers can have axial ratios (ratio of length to diameter) of 3000 or more.

Synthetic polymers have been developed that possess desirable characteristics, such as a high softening point to allow for ironing, high tensile strength, adequate stiffness, and desirable fabric qualities. These polymers are then formed into fibers with various characteristics.

Nylon (a generic term for polyamides) was developed in the 1930's and used for parachutes in World War II. This synthetic fiber, known for its strength, elasticity, toughness, and resistance to abrasion, has commercial applications including clothing and carpeting. Nylon has special properties which distinguish it from other materials. One such property is the elasticity. Nylon is very elastic, however after elastic limit has been exceeded the material will not return to its original shape. Like other synthetic fibers, Nylon has a large electrical resistance. This is the cause for the build-up of static charges in some articles of clothing and carpets.

From textiles to bullet-proof vests, fibers have become very important in modern life. As the technology of fiber processing expands, new generations of strong and light weight materials will be produced.

Words and Expressions

isoprene [ˈaɪsə,priːn] *n.* 异戊二烯
vulcanize [ˈvʌlkənaɪz] *v.* 硫化
sulfur [ˈsʌlfə] *n.* 硫黄
fragment [ˈfrægmənt] *n.* 链段
styrene [ˈstaɪriːn] *n.* 苯乙烯
butadiene [,bjuːtəˈdaɪiːn] *n.* 丁二烯
initiator [ɪˈnɪʃɪeɪtə] *n.* 引发剂
thermoplastic [,θɜːməˈplæstɪk] *adj.* 热塑性的；*n.* 热塑性塑料
thermoset [ˈθɜːməset] *adj.* 热固性的；*n.* 热固性塑料
insulation [,ɪnsjʊˈleɪʃən] *n.* 绝缘
tough [tʌf] *adj.* 韧性的
resilient [rɪˈzɪlɪənt] *n.* 弹性，有弹性；*adj.* 回弹的，反弹的
category [ˈkætɪɡərɪ] *n.* 种类
artificial [,ɑːtɪˈfɪʃəl] *adj.* 人工的，人造的
patent [ˈpeɪtənt, ˈpætənt] *v.* 取得专利权；*n.* 专利、许可证
launch [lɔːntʃ, lɑːntʃ] *v.* 使运动，送上轨道
nylon [ˈnaɪlən] *n.* 尼龙
polyester [ˈpɒlɪestə, ˌpɒlɪˈestə] *n.* 聚酯
rayon [ˈreɪɒn] *adj.* 人造丝的，人造纤维的；*n.* 人造丝，人造纤维
acrylic [əˈkrɪlɪk] *adj.* 丙烯酸的；*n.* 丙烯酸（类）树脂（=acrylic resin）
durability [,djʊərəˈbɪlɪtɪ] *n.* 持久性，耐久性
axial [ˈæksɪəl, ˈæksjəl] *adj.* 轴的，轴向的，轴心的
ironing [ˈaɪənɪŋ] *n.* 熨平
parachute [ˈpærəʃuːt] *n.* 降落伞

textile ['tekstaɪl]   *n*. 纺织品，织物
bullet-proof vest   防弹衣

Notes

1. Thermoplastics soften on heating and harden on cooling while thermoset, on heating, flow and cross-link to form rigid material which does not soften on future heating.

译文：热塑性塑料受热时软化，冷却后又可变硬；而热固性塑料在受热时熔融流动，并发生交联形成具有一定刚性的材料，但是再使它受热这个过程不可反复（即再加热不可再次软化）。

2. From textiles to bullet-proof vests, fibers have become very important in modern life. As the technology of fiber processing expands, new generations of strong and light weight materials will be produced.

译文：从普通织物到防弹衣，纤维在我们生活中变得越来越重要了。随着纤维加工技术的推广，将可生产出高强度而轻质的新性能纤维。

Reading Comprehension

1. What are the common types of rubber?
2. What are the applications of polyethylene?
3. Do you know what kind of man-made fiber? What can they do?

Reading Material

Polymers for Food Packing　聚合物在食品包装上的应用

Both thermosets and thermoplastics are used as food-contact materials, though thermoplastics predominate in this application. Examples of the former are phenol- and urea-formaldehyde, while probably the best known example of the latter is low-density poly(ethylene). Other linear polymers are used include high-density poly(ethylene), poly(propylene), and PVC, all of which find quite extensive use. Polymers for food packaging may be in the form of films and other flexible items, or in the form of rigid containers, such as clear drinks bottles or opaque cartons for dairy products.

For packaging, there are a variety of requirements. The foodstuff must be preserved for the customer until the moment of consumption in a form that is palatable and wholesome. This means the polymer will usually require good barrier properties, both to prevent loss of moisture from the food, and to prevent ingress of undesirable, possibly odorous contaminants from outside. In order to obtain optimum preservative properties, polymers are often employed in multilayers, each layer being a different polymer, so that the whole polymer system presents a barrier that is appropriately impervious. Unfortunately, blended films and used containers of this kind are difficult to recycle, and thus present a potential environmental problem.

It is critical that plastics used in food contact applications are not themselves a source of contamination of the food. High molecular weight polymers, because of their limited solubility in either organic or aqueous media, rarely if ever migrate into the food and hence present little toxic hazard. However, in order that polymers have the necessary properties of stability towards heat and light, as well as storage stability and flexibility, a variety of additives are blended into the polymer, and these do have the potential to migrate into the food, causing problems ranging from alteration of the taste to possible toxicity.

There is a considerable volume of legislation throughout the world governing what may be allowed to migrate from a polymeric packaging material into food in contact with it. Typically, within European Union legislation, no more than 10 mg/dm^3 of any component may be allowed to migrate into the food, a figure which is modified to 60 mg/kg of foodstuff for those items whose contact area cannot be readily measured.

Words and Expressions

phenol ['fiːnəl]   *n.* 苯酚
urea ['jʊərɪə]   *n.* 尿素
opaque [əʊ'peɪk]   *adj.* 不透明的
carton ['kɑːtən]   *n.* 纸板箱
palatable ['pælətəbl]   *adj.* 美味的
wholesome ['həʊlsəʊm]   *adj.* 合乎卫生的
additive ['ædɪtɪv]   *n.* 添加剂

 ## History, Inheritance and Development

Wanhua Chemical Attended the PU CHINA 2021 with its Life-cycle Sustainable Solutions for Polyurethane Materials

The 18th China International Polyurethane Exhibition, a.k.a. PU China 2021, was held in Shanghai from July 28-30. Wanhua Chemical presented the full life cycle sustainable solutions for polyurethane materials. Through the display of material solutions, the release of sustainable work reports and the display of recycled polyurethane products, we introduced sustainable solutions covering the full life cycle of polyurethane materials to the attendees.

During the exhibition, Wanhua Chemical presented detailed demonstrations on sustainable solutions of polyurethanes in multiple areas including home construction, recreation, and green energy. Under China's two carbon goals, say, "carbon-peak by 2030" and "carbon-neutrality by 2060", the polyurethane industry covers all aspects of energy saving and carbon reduction in social life. By developing low-carbon products, optimizing system formulations, and focusing on material recycling, Wanhua Chemical is providing high-performance raw materials for downstream enterprises while focusing on low-carbon environment protection across the entire industrial chain, boosting the green sustainability of the whole society.

It is worth mentioning that Wanhua Chemical exhibited the polyurethane rigid foam recycling solution for the first time in this exhibition. By chemically degrading the insulation layer in waste home appliances to make new foam products, it realizes the recycling of polyurethane rigid foams, thus reducing waste foam pollution and promoting circular economy.

From the material development to production, and from the product application to recycling, Wanhua Chemical stands ready to join hands with all relevant parties to promote the energy-saving and low-carbon process across the entire industry chain, and make every effort to achieve the two carbon goals!

Practice and Training

Synthesis of Super-absorbent Polymers

Super-absorbent polymers (SAPs) refer to a three-dimensional network polymer, water-swellable, water-insoluble, organic or inorganic material that can absorb thousands of times its own weight of distilled water. It is widely used in various fields, such as: agricultural, biomedical, daily physiological products, separation technology and wastewater treatment.

Polyaspartic acid (PASP) resin is a kind of biodegradable polymer material with high water absorbency on account of free carboxylic groups and amide groups in its molecular chains.

Nowadays, most PASP resin is prepared by chemical cross-linking reaction with a cross-linking agent.

The method of preparing PASP resin by chemical reaction is generally composed of three steps: they are cross-linking, hydrolysis, and drying.

PASP resin was prepared from PSI with a molecular mass of 12,000-199,500 Da which was synthesized by thermal polycondensation of Laspartic acid as described in the previous section. One gram of PSI was dissolved in 20 ml of an organic aprotic polar solvent such as DMF in a 500 ml beaker with magnetic stirring, and 10 ml deionized water as a dispersant was added into the beaker. The mixture containing PSI, DMF, and deionized water was stirred for 0.5 h, 0.08 g of cross-linking agent was added into the beaker, and the cross-linking reaction was carried out for 2 h at 40℃. The cross-linked PSI was prepared, and the imide ring of the cross-linked polymer was hydrolyzed with NaOH at 40℃ until the pH was 9. Then, 50 ml methanol was added to cause precipitation, and the precipitate was filtered and dried at 40℃ under a vacuum of 0.095MPa. Thus, a cross-linked PASP resin as a superabsorbent polymer was obtained.

Unit Four

Safety Engineering

安全工程

Lesson One A Brief History of Safety
安全工程简史

Of course, the need for safety has always been with us. One of the earliest written references to safety is from the Code of Hammurabi, around 1750 B.C. His code stated that if a house was built and it fell due to poor construction, killing the owner, then the builder himself would be put to death. The first laws covering compensation for injuries were codified in the Middle Ages.

Around 1834, Lloyd's Register of British and Foreign Snipping was created, institutionalizing the concept of safety and risk analysis. In 877 Massachusetts passed a law to safeguard machinery and also created employers' liability laws.

At the end of the 19th century, a rash of boilers exploding gave urgency and impetus to the American Society of Mechanical Engineers to create the Boiler and Pressure Vessel design codes and standards. Beginning in 1911, the United States saw safety groups forming, and the National Safety Council was founded in 1913.

Around the 1920s private companies started to create formalized safety programs. The early 1930s was the beginning of the implementation of accident prevention programs across the United States. By the end of the decade, the American National Standards Institute had published hundreds of industrial manuals.

Most of the current safety techniques and concepts were born at the end of World War II. Operations research led the way, suggesting that the scientific method could be applied to the safety profession. In fact, operations research gave some legitimacy to the use of quantitative analysis in predicting accidents.

However, the system safety concept and profession really started during the American military missile and nuclear programs in the 1950s and 1960s. Liquid-propellant missiles exploded frequently and unexpectedly. During that period the Atlas and Titan programs saw many missiles blow up in their silos during practice operations. Some of the accident investigations found that these failures were due to design problems, operations deficiencies, and poor management decisions.

Because of the loss of thousands of aircraft and pilots during the same time frame, the U.S. Air

Force started to pull together the concepts of system safety, and in April 1962 published BSD Exhibit 62-41, "System Safety Engineering for the Development of Air Force Ballistic Missiles."

Safety was also starting to enter the public mind. Ralph Nader publicized safety concerns during the mid-1960s and started making people aware of how dangerous cars really were with his book, *Unsafe at Any Speed* (published in 1965, Grossman, NY). He continued being a powerful voice to the U.S. Congress to bring automobile design under federal control and to regulate consumer protection.

In the United Kingdom in the early 1960s, Imperial Chemical Industries started developing the concept of the HAZOP study (a chemical industry safety analysis). In 1974 it was presented at an American Institute of Chemical Engineers conference on loss prevention.

The U.S. National Aeronautics and Space Administration (NASA) sponsored government-industry conferences in the late 1960s and early 1970s to address system safety. Part of this was safety technology transfer from the "man-rating" program-to develop ballistic missiles sage enough to carry humans into space-of the Mercury program.

In 1970 the Occupational Safety and Health Administration (OSHA) published industrial safety requirements. Later in the decade, the U.S. military published Mil-Std-882, "Requirements for System Safety Program for Systems and Associated Subsystems and Equipment." This document is still considered the cornerstone of the system safety profession. It is one of the most cited requirements in procurement contracts. Most of the safety analysis techniques were created during the heady days of safety from the 1950s to the 1980s.

OSHA published a process safety standard for hazardous materials in 1992. This is one of the strongest cross-fertilizations of system safety techniques taken from various industries and applied to the chemical industry. It is obvious that the system safety engineering profession, like all professions, has evolved overtime. In most cases, out of necessity an unacceptable number of deaths, accidents, and loss of revenue engineers have been forced to take a more serious approach to designing safety into both systems and products.

Words and Expressions

institutionalize [ˌɪnstɪˈtjuːʃənəlaɪz]	vt. 使制度化
compensation [ˌkɒmpenˈseɪʃən]	n. 补偿，赔偿，酬报
implementation [ˌɪmplɪmenˈteɪʃən]	n. 贯彻，实施，手段
legitimacy [lɪˈdʒɪtɪməsɪ]	n. 合法性
quantitative analysis	定量分析
investigation [ɪnˌvestɪˈgeɪʃən]	n. 调查，调查研究
cornerstone [ˈkɔːnəstəʊn]	n. 基础
procurement contracts	采购合同
evolve [ɪˈvɒlv]	vt. 发展，使演变
revenue [ˈrevɪnjuː]	n. 税收，收益，总收入
approach [əˈprəʊtʃ]	vt. or n. 向…靠近，接近，近似
Lloyd's Register	劳埃德船级社
institutionalize [ˌɪnstɪˈtjuːʃənəlaɪz]	v. 使制度化
Pressure Vessel	压力容器

Liquid-propellant　　　　　　　　　　　　　火箭引擎中的液体燃料
NASA　　　　　　　　　　　　　　　　　　美国国家航空航天局
OSHA　　　　　　　　　　　　　　　　　　职业安全与健康管理总署
Mil-Std-882　　　　　　　　　　　　　　　军标-882

Notes

1. At the end of the 19th century, a rash of boilers exploding gave urgency and impetus to the American Society of Mechanical Engineers to create the Boiler and Pressure Vessel design codes and standards.

句子分析：句子的结构为"主语+谓语+宾语"，谓语形式为 give sth. to sb.。

译文：19世纪末，大量的锅炉爆炸激发并促进了美国机械工程师协会创建了锅炉和压力容器设计规则和标准。

2. Operations research led the way, suggesting that the scientific method could be applied to the safety profession.

句子分析：that 引导从句为 suggesting 宾语从句。

译文：管理研究提出了方向表明科学的方法可以应用到安全职业中去。

3. Part of this was safety technology transfer from the "man-rating" program to develop ballistic missiles safe enough to carry humans into space-of the Mercury program.

句子分析：句子的结构为"主语+系词+表语"。

译文：其部分内容为安全技术从人身评价方案转向为发展弹道导弹足够安全地将人类带入太空的水星计划。

4. NASA：美国国家航空航天局（National Aeronautics and Space Administration）简称 NASA，是美国负责太空计划的政府机构。总部位于华盛顿哥伦比亚特区，拥有最先进的航空航天技术，它在载人空间飞行、航空学、空间科学等方面有很大的成就。它参与了包括美国阿波罗计划、航天飞机发射、太阳系探测等在内的航天工程。为人类探索太空做出了巨大的贡献。

5. OSHA：职业安全与健康管理局，该局主要设有国家办公室和区域办公室。国家办公室包括：合作和国家项目部（DCSP）、标准和指导部（DSG）、科学技术和医药部（DSTM）、执行项目部（DEP），以及评估和分析部（DEA）。按照行政区划不同，在全国范围共设置 10 个区域办公室，负责各区域的职业安全与健康管理。在区域办公室以下，还设有 73 个地区办公室，直接负责具体的执行任务。区域办公室及地区办公室都是联邦职业安全与健康管理局的派出机构。

Reading Comprehension

1. What is the earliest reference to safety according to the text?
2. When were the concept of safety and risk analysis institutionalized?
3. When were the most of the current safety techniques and concepts born at?
4. What led to the accident during that period the Atlas and Titan program?
5. What is considered the cornerstone of the system safety profession?

Reading Material

Why Do We Need Safety Engineering? 为什么需要安全工程

It is difficult to open a newspaper or turn on the television and not be reminded how dangerous our world is. Both large-scale natural and man-made disasters seem to occur on an almost daily basis. An accident at a plant in Bhopal, India, killed over 2,500 people. A nuclear power plant in the Ukraine exploded and burned out of control, sending a radioactive cloud to over 20 countries, severely affecting its immediate neighbors' live-stock and farming.

A total of 6.7 million injuries and illnesses in the United States were reported by private industry in 1993. Two commuter trains in metropolitan Washington, D.C., collided in 1996, killing numerous passengers. Large oil tankers ran aground in Alaska and Mexico, spilling millions of gallons of oil and despoiling the coastline. An automobile air-bag manufacturing plant exploded, killing one worker, after it had had over 21 fire emergencies in one year. Swarms of helicopters with television cameras were drawn to the plant after every call, creating a public relations nightmare and forcing the government to shut down the plant temporarily.

An airliner crashed into an apartment building in downtown Sao Paolo, Brazil, killing all on board and many in the apartment building. Another, airplane mysteriously dipped and spun into the ground in Sioux City, Iowa. Two airplanes collided on a runway in the Philippines. An airliner crashed into the Florida Everglades after an oxygen generator exploded in the cargo hold, killing all 110 people on board.

In 1995 the Fremont, California, Air Route Traffic Control Center lost power, causing radar screens covering northern California, western Nevada, and 18 million square miles of Pacific Ocean to go dark for 34 minutes while 70 planes were in the air, almost resulting in two separate midair collisions. In another incident, a worker in downtown Chicago cut into a cable and brought down the entire Air Route Traffic Control System for thousands of square miles.

Some of these accidents occurred many years ago, some of them occurred very recently. Many of the accidents crossed international borders and affected millions of people in other countries. Many more did not extend beyond national borders but still affected a great number of people. And some of the accidents didn't kill anyone.

We all know how quickly technology is changing; as engineers, it is difficult just to keep up. As technology advances by leaps and bounds, and business competition heats up with the internationalization of the economy, turnaround time from product design to market launch is shrinking quickly. The problem quickly becomes evident; how do we build products with high quality, cheaply, quickly and still safely?

An American Society of Mechanical Engineers national survey found that most design engineers were very aware of the importance of safety and product liability in designs but did not know how to use the system safety tools available. In fact, most of the engineers who responded said that the only safety analyses they used were the application of safety factors in design, safety checklists, and the use of compliance standards. Almost 80 percent of the engineers had never taken a safety course in college, and more than 60 percent had never taken a short course in safety through work. Also, 80 percent had never attended a safety conference and 70 percent had never attended a safety lecture.

So, how do engineers design, build, and operate systems safely if they have never really been prepared for it? And, to make matters worse, engineers are now more frequently called to testify in court about failures in their designs.

Like most engineering problems, this one does have a solution. And the solution is not that difficult to implement, nor costly. What it does entail is considerable forethought and systematic engineering analysis. Of course, system of safety engineering is not difficult to apply. In fact, it is almost easy.

Words and Expressions

explode [ɪksˈpləʊd]   vt. 使爆炸，破除，戳穿
radioactive cloud   放射云
live-stock   家畜，牲畜
commuter [kəˈmjuːtə]   n. 乘公交车上下班者
despoil [dɪsˈpɔɪl]   vt. 抢劫，掠夺
coastline [ˈkəʊstlaɪn]   n. 海岸线
helicopter [ˈhelɪkɒptə]   n. 直升机
collide [kəˈlaɪd]   vi. or vt. 碰撞，互撞
cargo [ˈkɑːgəʊ]   n. 货物，一批（货物）
entail [ɪnˈteɪl]   v. 使必需，使承担

Lesson Two   The System Safety Process
系统安全过程

The system safety process is really an easy concept to grasp. The overall purpose is to identify hazards, eliminate or control them, and mitigate the residual risks. The process should combine management oversight and engineering analyses to provide a comprehensive, systematic approach to managing the system risks. Fig. 4-1 details this process.

As with any problem, the first step is to define the boundary conditions or analysis objectives. That is the scope or level of protection desired. One must understand what level of safety is desired at what cost. The engineer needs to answer the question: how safe is safe enough? Other questions to ask are:

What constitutes a catastrophic accident?

What constitutes a critical accident?

Is the cost of preventing the accident acceptable?

Most industries approach this step in the same way. However, how they differentiate among catastrophic, critical, and negligible hazards may vary. The engineering will need to modify the definitions to fit the particular problem. What is important is that these definitions are determined before work begins. A rule-of-thumb definition for each is: Catastrophic any event that may cause death or serious personnel injury, or loss of system (e.g. anhydrous ammonia tanker truck overturns, resulting in a major spill). Critical-any event that may cause severe injury, or loss of mission-critical hardware or high-dollar-value equipment(e.g. regulator fails open and over-pressurizes a remote hydraulic line, fig. 4-1 the system safety process damaging equipment and bringing the system down for some days).

Minor-any event that may cause minor injury or minor system damage, but does not significantly impact the mission (e.g. pressure control valve fails open, causing pressure drops and increased caustic levels).

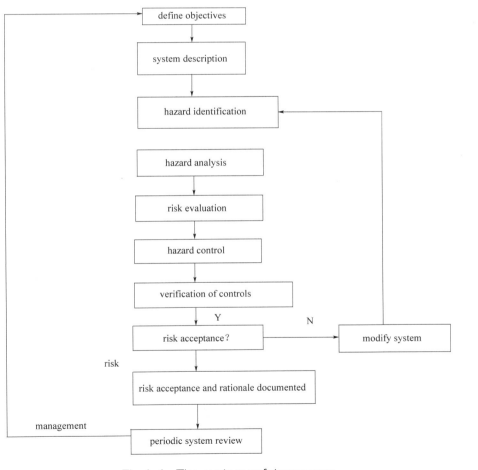

Fig.4-1　The system safety process

Negligible-any event that does not result in injury or system damage and does not affect the mission (e.g. lose commercial power, causing shutdown of plant cafeteria). The next step is system description. Some time should be given to grasping how the system works and how the hardware, software, people, and environment all interact. If the system is not described accurately, then the safety analysis and control program will be flawed.

1. Hazard Identification

Hazard identification is a crucial part of the system safety process. It really is impossible to safeguard a system or control risks adequately without first identifying the hazards. An all-too-frequent mistake in safety engineering is to skip over this step, or not to give it adequate attention. The hazard identification process is a kind of "safety brainstorming". The purpose is to identify as many hazards as are possible and credible. Through this process the engineer develops a preliminary hazard list (PHL) and later will assess the impact on the system.

To develop a PHL the engineer will want to use various methods to gather the most exhaustive list possible. This may include:

Survey the site;

Interview site personnel;

Convene a technical expert panel;

Analyze and compare similar systems;

Identify codes, standards, and regulations;

Review relevant technical data (electrical and mechanical drawings, analyses, operator and manuals and procedures, engineering reports, etc.);

Analyze energy sources (voltage/current sources, high/low temperature sources, etc.).

The next step is to analyze the hazards identified. A hazard analysis is a technique for studying the cause/consequence relation of the hazard potential in a system. The purpose is to take the preliminary hazard list one level deeper and assess how each hazard affects the system. Is it catastrophic? Or is it critical? The hazard analysis will also assist the engineer in further assessing which hazards are important and which are not and therefore do not need further study. There are various hazard analysis techniques that are commonly used in different industries.

After hazards have been identified and analyzed, the engineer needs to control their occurrence or mitigate their effects. This is done by evaluating the risks. Is the hazard likely to occur? If it does, how much damage will result from the incident? The engineer needs to understand the relationship between hazard cause and effect. With this information, the associated risks are then ranked and engineering management is better able to determine which risks are worth controlling and which risks require less attention.

2. Hazard Control

After evaluating the risks and ranking their importance, the engineer must control their effects. Controls fall into two broad categories: engineering controls and management controls. Engineering controls are changes in the hardware that either eliminate the hazards or mitigate their risks. Some example engineering controls include: adding a relief valve to a 2,000-psi oxygen system; building a berm around an oil storage tank; using only hermetically sealed switches in an explosive environment; or putting in hard stops in rotating machinery to prevent over-torquing.

Management controls are changes made to the organization itself. Developing and implementing a plant safety plan is a good method of applying management controls to hazards. Some examples are: using production line employees as safety representatives for their areas; requiring middle-management reviews and approvals of any plant or system modifications to consider safety implication; or assigning signature authority to safety engineers for all engineering change orders and drawings.

Once controls are in place, a method needs to be used to verify that the controls actually control the hazards or mitigate the risks to an acceptable level. Verification of hazard controls is usually accomplished through the company or engineering management structure. The most frequent means is inspection. However, as we all know, inspection is also one of the most expensive ways to assure that controls are in place. An effective method of hazard control verification is the use of a closed-loop tracking and resolution process.

3. Risk Acceptance

Safety is only as important as management wants to make it. At this point in the safety process this becomes obvious. After the system has been studied and hazards identified, then analyzed and evaluated with controls in place, management must make the formal decisions of which risks they are willing to take and which ones they will not take. At this point a good cost benefit analysis will help management make that decision. Sometimes this is not easy.

Part of the risk acceptance process is a methodical decision making approach. If the risks are not acceptable, then the system must be modified and the hazard identification process must be followed once again. If the risks are acceptable, then good documentation with written rationale is imperative to protect against liability claims.

Probably one of the key points of the system safety process is that it is a closed loop system. This means that the engineering and management organizations periodically review the safety program, engineering processes, management organizations, and product field use. The American automobile industry has lost billions of dollars in automobile recalls due to safety problems, some of which possibly could have been avoided by periodic review of product use.

Words and Expressions

mitigate ['mɪtɪgeɪt]   *vt.* 使缓和，减轻
differentiate [ˌdɪfə'renʃɪeɪt]   *v.* 区别，区分
catastrophic [ˌkætə'strɒfɪk]   *adj.* 悲惨的，灾难的
critical ['krɪtɪkəl]   *adj.* 批评的，严谨的
minor ['maɪnə]   *adj.* 较小的，较次要的
negligible ['neglɪdʒəbl]   *adj.* 无关紧要的
rule-of-thumb   单凭经验的方法
anhydrous ammonia   无水氨
hydraulic [haɪ'drɔːlɪk]   *adj.* 水力的，水压的

caustic ['kɔːstɪk]	*adj.* 腐蚀性的
flaw [flɔː]	*n.* 裂纹，有瑕疵
all-too-frequent	太频繁
preliminary [prɪ'lɪmɪnərɪ]	*adj.* 预备的，初步的
exhaustive [ɪɡ'zɔːstɪv]	*adj.* 耗尽的，枯竭的
convene [kən'viːn]	*vt.* 召集，集合
relevant ['relɪvənt]	*adj.* 信赖的，依靠的
code [kəʊd]	*n.* 法律，法规
mitigate ['mɪtɪɡeɪt]	*v.* 减轻
hermetically [hɜː'metɪkəlɪ]	*adv.* 密封地
closed-loop	闭环
rationale [ˌræʃə'nɑːlɪ]	*n.* 基本原理
imperative [ɪm'perətɪv]	*adj.* 必要的，急迫的

Notes

1. One must understand what level of safety is desired at what cost.

句子的结构为：主语+谓语+宾语从句的形式。

译文：人们必须明白何种等级的安全需要何等的代价。

2. To develop a PHL the engineer will want to use various methods to gather the most exhaustive list possible.

句子分析：句中不定式"to develop…"作主语，"to gather…"作目的状语。

译文：要完善预期危险列表，工程人员要使用各种方法来搜集最详尽的列表。

3. With this information, the associated risks are then ranked and engineering management is better able to determine which risks are worth controlling and which risks require less attention.

译文：依据此类信息，可以列出相关的风险，且工程管理时最好能确认哪类风险值得重视而哪类不需要过度担心。

4. After the system has been studied and hazards identified, then analyzed and evaluated with controls in place, management must make the formal decisions of which risks they are willing to take and which ones they will not take.

译文：在对系统进行了研究并识别了风险，然后在适当的控制下进行了分析和评估之后，管理层必须正式决定他们愿意承担哪些风险，不愿意承担哪些风险。

Reading Comprehension

1. What is the overall purpose of the system safety process?

2. What questions does the engineer need to answer when he understand the level of safety?

3. What are the rule-of-thumb definitions of catastrophic, critical, minor, and negligible hazards?

4. What are included in gathering the most exhaustive list possible?

5. Please give some examples to interpret what the engineering controls include.

Reading Material

The Happening of an Accident 事故的发生

We may all say accidents happen. However, their occurrence may not only take human lives, destroy millions of dollars in property and lost business, they may also cost us our jobs and reputations. The Bhopal, India, accident in 1984 released methyl isocyanate and caused over 2,500 fatalities. In 1986, the NASA Space shuttle challenger disintegrated in flight in front of millions of television viewers and killed sever astronauts, brought NASA to a standstill for two years, and cost the agency billions of dollars. A petroleum refinery blew up in Houston, Texas, in 1989, killing 23 workers, damaging property totaling US ＄750 million, and spewing debris from the explosion over an area of 9kin. Many thought that after the Three mile Island and Chernobyl nuclear power plant disasters we would finally get a handle on how to prevent accident. U.S. government statistics indicate that more than 350 chemical accidents a year result in death, injury or evacuation. In 1991 and 1992 fifteen major petrochemical accidents destroyed more than ＄1 billion in property.

Accidents don't just happen; they are a result of a long process, with many steps. Many times all of these steps have to be completed before an accident can occur. If the engineer can prevent one or more of these accident steps from occurring, then he can either prevent the mishap of at least mitigate its effects. Part of system safety strategy is to intervene at various points along that accident time line.

An accident is an unplanned process of events that leads to undesired injury, loss of life, damage to the system or the environment. This means that death in war is no accident, but a jeep crashing on the way to battle is.

An incident or near-miss is an almost-accident. Three mile Island was a radioactive near-miss. No massive quantities of radioactivity were released to the environs, but they almost were. Figure 1 shows the events that lead to an accident.

An incident or near-miss is an almost-accident. Three Mile Island was a radioactive near-miss. No massive quantities of radioactivity were released to the environs, but they almost were. Fig. 4-2 shows the events that lead to an accident.

Preliminary events can be anything that influences the initiating event. Examples of preliminary events could be long working hours for chemical plant operators or poor or incomplete pump maintenance. Preliminary events set the stage for a hazardous condition. If we can eliminate the preliminary events or hazardous condition, then the accident cannot advance to the next step-initiating events.

The initiating event, sometimes called the trigger event, is the actual mechanism or condition that causes the accident to occur. It can be thought of as the spark that lights the fire. For example, a valve sticks open a process feed line, an electrical short causes a spark at a fueling depot, a pressure regulator fails open in a cryogenic system or a 220-V power feed is mated with a 110-V system.

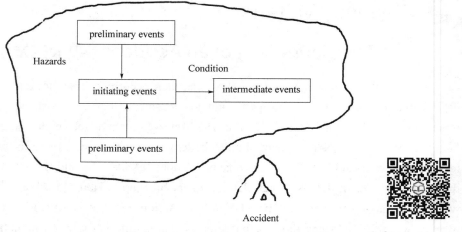

Fig.4-2　Events that lead to an accident

Safety accident case

　　Intermediate events can have two effects. They may propagate or ameliorate the accident. Functioning relief valves in a pressure system will ameliorate a system over pressurization. No pressure relief will propagate the hazardous condition and create an accident of system pressure rupture. Defensive driving on highways helps us protect ourselves from the "other" crazy driver or ameliorate the effects of his bad driving. Obviously, drunk driving does the opposite, propagating and intensifying an already dangerous situation.

Words and Expressions

methyl isocyanate	甲基异氰酸盐
fatality [fə'tælɪtɪ]	n. 灾祸
petroleum refinery	炼油厂
spewing debris	压榨碎片
evacuation [ɪˌvækjʊ'eɪʃən]	n. 撤离
mishap ['mɪshæp, mɪs'hæp]	n. 灾祸
mitigate ['mɪtɪgeɪt]	vt. 减轻，使缓和
time line	年表，活动时间表
trigger ['trɪgə]	vt. 出发，激发
cryogenic [ˌkraɪə'dʒɛnɪk]	adj. 低温的
ameliorate [ə'miːlɪəreɪt]	v. 改善，改进
pressurization [ˌpreʃəraɪ'zeɪʃn]	n. 加压
propagate ['prɒpəgeɪt]	v. 宣传，传播
rupture ['rʌptʃə(r)]	v. 破裂
ensue [ɪn'sjuː]	vi. 跟着发生，继起
defensive [dɪ'fensɪv]	adj. 防御的，防卫的
crazy ['kreɪzɪ]	adj. 疯狂的，发狂的

Lesson Three Basic Principles for Controlling Chemical Hazards
管理化学危险的基本原则

Chemicals are considered highly hazardous for many reasons. They may cause cancer, birth defects, induce genetic damage, and cause miscarriage, or otherwise interfere with the reproductive process. Or they may be a cholinesterase inhibitor, a cyanide, or other highly toxic chemical that, after a comparatively small exposure, can lead to serious injury or even death. Working with compounds like these generally necessitates implementation of additional safety precautions.

The goal of defining precisely, in measurable terms, every possible health effect that may occur in the workplace as a result of chemical exposures cannot realistically be accomplished. This does not negate the need for laboratory personnel to know about the possible effects as well as the physical hazards of the hazardous chemicals they use, and to protect themselves from these effects and hazards. Controlling possible hazards may require the application of engineering hazard controls (substitution, minimization, isolation, ventilation) supplemented by administrative hazard controls (planning, information and training, written policies and procedures, safe work practices, and environmental and medical). Personal protective equipment (e.g. gloves, goggles, coats, respirators) may need to be considered if engineering and administrative controls are not three will be necessary to control the hazards.

It should also be kept in mind that the risks associated with the possession and use of a hazardous chemical are dependent upon a multitude of factors, all of which must be considered before acquiring and using a hazardous chemical. Important elements to examine and address include: the knowledge of and commitment to safe laboratory practices of all who handle the chemical; its physical, chemical, and biological properties and those of its derivatives; the quantity received and the manner in which it is stored and distributed; the manner in which it is used; the manner of disposal of the substance and its derivatives; the length of time it is on the premises, and the number of persons who work in the area and have open access to the substance (the Preliminary Chemical Hazard Assessment Form can be used as part of this assessment). The decision to procure a specific quantity of a specific hazardous chemical is a commitment to handle it responsibly from receipt to ultimate disposal.

1. Basic Hazard Control Rational

The basic principles for controlling chemical hazards can be broken down into three broad categories: engineering controls, administrative controls, and personal protective equipment. Hazards must be controlled first by the application of engineering controls that are supplemented by administrative controls. Personal protective equipment is only considered when other controls are not technically, operationally or financially feasible. Typically, combinations of all methods are necessary in controlling chemical hazards.

2. Engineering Hazard Controls

Engineering hazard controls may be defined as an installation of equipment, or other physical facilities including, if necessary, the selection and arrangement of experimental equipment. Engineering controls remove the hazard, either by initial design specifications or by applying methods of substitution, minimization, isolation, or ventilation. Engineering controls are the most effective hazard control methods, especially when introduced at the conceptual stage of planning when control measures can be integrated more readily into the design. They tend to be more effective than other hazard controls (administrative controls and personal protective equipment) because they remove the source of the hazard or reduce the hazard rather than lessen the damage that may result from the hazard. They are also less dependent on the chemical user who, unfortunately, is subjected to all of the frailties which befall.

Substitution refers to the replacement of a hazardous material or process with one that is less hazardous (e.g. the replacement of mercury thermometers with alcohol thermometers or dip coating materials rather than spray coating to reduce the inhalation hazard). Minimization is the expression used when a hazard lessened by the hazardous process. Hence, the quantity of hazardous materials used and stored is reduced, lessening the potential hazards. Isolation is the term applied when a barrier is interposed between a material, equipment or process hazard and the property or persons who might be affected by the hazard.

Ventilation is used to control toxic and/or flammable atmospheres by exhausting or supplying air to either remove hazardous atmospheres at their source or dilute them to a safe level. The two types of ventilation are typically termed local exhaust and general ventilation. Local exhaust attempts to enclose the material, equipment or process as much as possible and to withdraw air from the physical enclosure at a rate sufficient to assure that the direction of air movement at all openings is always into the enclosure. General ventilation attempts to control hazardous atmosphere by diluting the atmosphere to a safe level by either exhausting or supplying air to the general area.

Local exhaust is always the preferable ventilation method but is typically more costly. For some situations, general ventilation may suffice but only if the following criteria are met: only small quantities of air contaminants are released into the area at fairly uniform rates; there is sufficient air movement between the person and contaminant source to allow sufficient air movement to dilute the contaminant to a safe level; only materials of low toxicity or flammability are being used; there is no need to collect or filter the contaminant before the exhaust air is discharged into the environment (including the rest of the building), and the contaminant will not produce or other damage to equipment in the area or in any way affect other building occupants outside the general use area.

3. Administrative Hazard Controls

All of the aforementioned engineering hazard control methods, in order to exist or be effective, require the application of "administrative hazard control" as either supplemental hazard controls or to ensure that engineering controls are developed, maintained, and properly functioning. Administrative hazard controls consist of managerial efforts to reduce hazards through planning, information and training (e.g. hazard communication), written policies and procedures (e.g. the Chemical Hygiene Plan), safe work practices, and environmental and medical surveillance (e.g. work place inspections, equipment preventive maintenance, and exposure monitoring). Because they primarily address the human element of hazard controls, they are of vital importance and are always used to control chemical hazards.

4. Personal Protective Equipment

As was mentioned earlier, when adequate engineering and administrative hazard controls are not technically, operationally, or financially feasible, personal protective equipment must be considered as a supplement. "Personal protective equipment" (PPE) includes a wide variety of items worn by an individual to isolate the person from chemical hazards. PPE includes articles to protect the eyes, skin, and the respiratory tract. PPE may be the only reasonable hazard control option, but for many reasons it is the least desirable means of controlling chemical hazards. PPE users must be aware of, and compensate for these undesirable qualities. PPE does not eliminate hazards but merely minimizes damage from hazards. The effectiveness of PPE is highly dependent on the user. PPE is oftentimes cumbersome and uncomfortable to wear. Each type of PPE has specific applications, advantages, limitations, and potential problems associated with their misuse and those using PPE must be fully knowledgeable of these considerations. PPE must match the hazards and the conditions of use and be properly maintained in order to be effective. Their misuse may directly or indirectly contribute to the hazard or create a new one. The material of construction must be compatible with the chemical's hazards and must maximize protection, dexterity, and comfort.

5. Every Hazard Can Be Controlled

Not all the previously mentioned principles are applicable to controlling the hazards of every chemical, but all chemical hazards can be controlled by the application of at least one of these methods. Ingenuity, experience, and a complete understanding of the circumstances surrounding the control problem will be required in choosing methods which will not only provide adequate hazard control, but which will consider development, installation, and/or operating costs as well as human factors such as user acceptance, convenience, comfort, etc.

Words and Expressions

cholinesterase [ˌkoʊləˈnestəˌreɪs]	n.	胆碱酯酶
cyanide [ˈsaɪənaɪd]	n.	氰化物
necessitate [nɪˈsesɪteɪt]	vt.	使成为必需
surveillance [sɜːˈveɪləns]	n.	监测，监督
commitment [kəˈmɪtmənt]	n.	承诺，应允的义务
derivative [dɪˈrɪvətɪv]	n.	衍生物
disposal [dɪsˈpəʊzəl]	n.	处置，处理，处理方式
facility [fəˈsɪlɪtɪ]	n.	设施，设备
frailty [ˈfreɪltɪ]	n.	意志薄弱，性格缺陷
befall [bɪˈfɔːl]	vt.	落到……的身上，降临于
preoccupation [priːˌɒkjʊˈpeɪʃən]	n.	偏见，成见
mercury thermometer		水银温度计
scale down		降低，减小
interpose [ˌɪntɜːˈpəʊz]	vt.	放入，插入

enclose [ɪnˈkləʊz]	vt.	密封，围包住
contaminant [kənˈtæmənənt]	n.	污染物
corrosion [kəˈrəʊʒən]	n.	腐蚀，侵蚀
cumbersome [ˈkʌmbəsəm]	adj.	麻烦的，笨重的
dexterity [dɛkˈstɛrɪti]	n.	（手）灵巧，熟练
ingenuity [ˌɪndʒɪˈnjuːɪtɪ]	n.	机灵，独创性，精巧

Notes

1. It should also be kept in mind that the risks associated with the possession and use of a hazardous chemical are dependent upon a multitude of factors, all of which must be considered before acquiring and using a hazardous chemical.

译文：人们还应该牢记的是风险是和财富联系在一起的，且危险化学品的使用取决于多种因素，所有这些必须在购买和使用危险化学品之前就要认识到。

2. The basic principles for controlling chemical hazards can be broken down into three broad categories: engineering controls, administrative controls, and personal protective equipment.

译文：控制化学品危险的基本原则可以分解为三个主要方面：工程控制，管理控制和个人防护设备。

3. Engineering hazard controls may be defined as an installation of equipment, or other physical facilities including, if necessary, the selection and arrangement of experimental equipment.

译文：工程危险控制或者可以定义为设备或其他物理机械的安装，如果需要的话还包括选择、安装实验设备。

4. As was mentioned earlier, when adequate engineering and administrative hazard controls are not technically, operationally, or financially feasible, personal protective equipment must be considered as a supplement.

译文：正如之前提到的，当合适的工程和管理危险控制不是在技术上、操作上或财物上可以实现的时候，个人防护设备就成了必需品。

Reading Comprehension

1. What may Controlling possible hazards require?
2. What kinds are the basic principles broken down into?
3. What is the definition of the Engineering hazard controls?
4. What does the "Personal protective equipment" (PPE) include?
5. What are engineering hazard controls, administrative hazard controls and personal protective equipment included respectively?

Reading Material

Machinery Equipment Safety　机械设备安全

The use of machinery at work carries a number of different risks, including hazards associated with the action the component of machine, its integrity or its operation. Effective machinery safety can be achieved through better design, improved layout and management procedures, including the selection and training of operators.

Many accidents in the workplace arise out of the use of machinery and tools ranging from hand tools through machine tools such as lathes and presses to large items of mobile mechanical plant e.g. cranes or fork lift trucks. Injuries may be caused in many ways such as:

- Articles dropped or falling e.g. from lifting plant;
- Contact with moving parts of machinery;
- Trapping, particularly of hands;
- Electric shocks, burns or scalds;
- Flying debris following fragmentation of machinery or materials being worked on;
- Emission of harmful substances from machinery.

Accidents not only cause human suffering, they also cost money, for example in lost working hours, training temporary staff, **insurance premiums**, fines and managers' time. By using safe, well-maintained equipment operated by adequately trained staff, you can help prevent accidents and reduce these personal and financial costs.

"Work equipment" is almost any equipment used by a worker at work including: machines such as circular saws, drilling machines, photocopiers, mowing machines, tractors, dumper trucks and power presses; hand tools such as screwdrivers, knives, hand saws and meat cleavers; lifting equipment such as lift trucks, elevating work platforms, vehicle hoists, lifting slings and bath lifts; other equipment such as ladders and water pressure cleaners.

1. What Risks Are There from Using Work Equipment?

Many things can cause a risk, for example:

- Using the wrong equipment for the job, e.g. ladders instead of access towers for an extended job at high level;
- Not fitting adequate guards on machines, leading to accidents caused by entanglement, shearing crushing, trapping or cutting;
- Not fitting adequate controls, or the wrong type of controls, so that equipment cannot be turned off quickly and safely, or starts accidentally;
- Not properly maintaining guards, safety devices, controls, etc. so that machines or equipment become unsafe;
- Not providing the right information, instruction and training for those using the equipment;
- Not fitting roll-over protective structures(ROPS) and seat belts on mobile work equipment where there is a risk of roll over (Note: this does not apply to quad bikes);
- Not maintaining work equipment or carrying out regular inspections and thorough examinations;
- Not providing the personal protective equipment needed to use certain machines safely, e.g. chainsaws, angle grinders.

2. How to Reduce the Risks

(1) Use the right equipment for the job

Many accidents happen because people have not chosen the right equipment for the work to be done. Controlling the risk often means planning ahead and ensuring that suitable equipment or machinery is available.

(2) Make sure machinery is safe

You should check the machinery is suitable for the work—think about how and where it will be used. All new machinery should be:

- CE marked;
- Safe-never rely exclusively on the CE mark to guarantee machinery is safe. It is only a claim by the manufacture that the equipment is safe. You must make your own safety checks;
- Provided with an EC Declaration of Conformity (ask for a copy if you have not been given one);
- Provided with instructions in English.

If you think that machinery you have bought is not safe, DO NOT USE IT.

Contact the manufacturer to discuss your concerns and if they are not helpful, contact your local HSE office for advice. Remember, it is your responsibility as an employer or a subcontractor to ensure you do not expose others to risk.

(3) Guard dangerous parts of machines

Controlling the risk often means guarding the parts of machines and equipment that could cause injury. Remember: use fixed guards wherever possible, properly fastened in place with screws or nuts and bolts which need tools to remove them; if employees need regular guard access to parts of the machine and a fixed guard is not possible, use an interlocked guard for those parts. This will ensure that the machine cannot start before the guard is closed and will stop if the guard is opened while the machine is operating, in some cases, e.g. on guillotines, devices such as photoelectric systems or automatic guards may be use and not easy to defeat, otherwise they may need modifying; make sure the guards allow the machine to be cleaned and maintained safely; where guards cannot give full protection, use jigs, holders, push sticks etc. to move the work piece.

(4) Make sure machinery and equipment are maintained in a safe condition

To control the risk you should carry out regular maintenance and preventive checks, and inspections where there is a significant risk. Some types of equipment are also required by law to be thoroughly examined by a competent person. Inspections should be carried out by a competent person at regular intervals to make sure the equipment is safe to operate. The intervals between inspections will depend on the type of equipment, how often it is used and environmental conditions. Inspections should always be carried out before the equipment is used for the first time or after major repairs. Keep a record of inspections made as this can provide useful information for maintenance workers planning maintenance activities.

- Make sure the guards and other safety devices (e.g. photoelectric systems) are routinely checked and kept in working order. They should also be checked after any repairs or modifications by a competent person.
- Check what the manufacture's instructions say about maintenance to ensure it is carried out where necessary and to the correct standard.
- Routine daily and weekly checks may be necessary, e.g. fluid levels, sures, brake function, guards.

When you enter a contract to hire equipment, particularly a long-term one, you will need to discuss what routine maintenance is needed and who will carry it out.

• Some equipment, e.g. a crane, needs preventive maintenance (servicing) so that it does not become unsafe.

• Lifting equipment, pressure systems and power presses should be thoroughly examined by a competent person at regular intervals specified in law or according to an examination scheme drawn up by a competent person. Your insurance company may be able to advise on who would be suitable to give you this help.

(5) Carry out maintenance work safety

Many accidents occur during maintenance work. Controlling the risk means following safe working practices, for example:

• Where possible, carry out maintenance with the power to the equipment off and ideally disconnected or with the fuses or keys removed, particularly where access to dangerous parts will be needed;

• Isolate equipment and pipelines containing pressurized fluid, gas, steam or hazardous material. Isolating valves should be locked off and the system depressurized where possible, particularly if access to dangerous parts will be needed.

• Support parts of equipment which could fall;

• Allow moving equipment to stop;

• Allow components which operate at high temperatures time to cool;

• Switch off the engine of mobile equipment, put the gearbox in neutral, apply the brake and, where necessary, chock the wheels;

• To prevent fire and explosions, thoroughly clean vessels that have contained flammable solids, liquid, gases or dusts and check them before hot work is carried out. Even small amounts of flammable material can give off enough vapor to create an explosive air mixture which could be ignited by a hand lamp or cutting/welding torch;

• Where maintenance work has to be carried out at height, ensure that a safe and secure means of access is provided which is suitable for the nature, duration an frequency of the task.

3. The Precaution Means in Practice

Accidents using the following equipment are common in small firms, but they can be prevented by some simple rules.

• Drilling machines

As with other cutting machines, the operator must be protected from the rotating chuck and swarf that is produced by the drill bit. Specially designed shields can be attached to the quill and used to protect this area. A telescoping portion of the shield can retract as the drill bit comes down into the workpiece. On larger bang or radial drills, a more universal type shield is typically applied.

• Lathes

There are three main mechanical safety considerations for lathes (engine, turret, etc.). One is the rotating chuck that could catch the operator's clothing, jewelry, hair, or hand and pull them into the machine. Two is the hazardous flying chips and coolant splash that are generated at the point of operation (where the tool contacts the workpiece being machined). The last safety consideration is a chuck wrench left in the chuck. To protect these areas, two shields can be applied—one around a portion of the chuck and the other at the point of operation. Larger sliding shields can protect both areas, providing the workpiece is not too long. On VRLs (vertical turret lathes), the safety concern is the rotating table and the point-of-operation swarf. Special barriers may have to be fabricated around the tables of these machines; shields can be provided at the point of operation.

- Milling machines

The main mechanical safety consideration for milling machines is the swarf that is generating at the point of operation. Another safety concern is the tool cutter, which could catch operator's clothing, jewelry, hair, or any other part of the body. Usually on smaller mills, the operator and other employees in the machine area are protected by shields. These shields can be applied around the perimeter of the table or bed area or close to the cutter, depending on the size of workpiece and the application. On larger milling machines, operators are sometimes protected by location; however, when working close to a cutting tool they must be protected from swarf.

- Grinding machines

Shields are usually applied applied to grinding machines to protect the operator from chips (swarf), sparks, coolant, or lubricant. A vacuum pedestal is also available to capture discharged debris. Other safety concerns for grinders are the adjustment of the work rests and the adjustable tongues or ends of the peripheral members at the top of each wheel. Work rests shall be kept adjusted closed to the wheel with a maximun opening of 1/8 in. The distance between the wheel periphery and the adjustable tongue or the end of the peripheral member at the top shall never exceed 1/4 in.

Words and Expressions

lathe [leɪð]	n.	车床
crane [kreɪn]	n.	起重机
fork lift truck		叉车
insurance premium		保险费
circular saw		圆锯
photocopier [ˈfəʊtəʊkɒpɪə(r)]	n.	复印机
mowing machine		割草机
screwdriver [ˈskruːˌdraɪvə-]	n.	螺丝起子
hand saw		手锯
vehicle hoist		升车机
lifting sling		升降索套
entanglement [ɪnˈtæŋglmənt]	n.	纠缠，缠结
safety device		安全防护装置
ROPS		翻车安全保护装置

angle device	安全防护装置
CE marked	CE 认证标志
HSE	健康、安全、环境管理体系的简称
Declaration of Conformity	符合性声明
guillotine ['gɪləti:n, gɪlə'ti:n]	n. 剪床
isolating valve	隔离阀
drilling machine	钻孔机
swarf [swɔ:f]	n. 切屑
VTL（vertical turret lathe）	立式转塔车床
milling machine	铣床
workpiece ['wɜ:kpi:s]	n. 工件
grinding machine	磨床
coolant ['ku:lənt]	n. 冷冻剂
lubricant ['lu:brɪkənt]	n. 润滑物，润滑油，润滑剂
vacuum ['vækjʊəm]	n. 真空，空间
vacuum pedestal	真空助力器座

 # History, Inheritance and Development

China Unveils Plan to Strengthen Safe Production of Hazardous Chemicals

BEIJING, March 21 (Xinhua) -- China has issued a plan to enhance work on the safe production of hazardous chemicals to prevent major accidents and improve workplace safety.

The plan, issued by the Ministry of Emergency Management, detailed the major targets to ensure the safe production of hazardous chemicals and fireworks, as well as oil and gas over the 14th Five-Year Plan period (2021—2025).

By 2025, major and serious accidents caused by hazardous chemicals will be effectively curbed, and a system for screening, prevention and control of hidden dangers of hazardous chemicals will be established, the plan said.

The monitoring and early warning system of safety risks will be further improved by 2025, and the intelligent control platforms for chemical parks, hazardous chemical firms, and large oil and gas storage bases will be constantly upgraded.

Looking to 2035, a clear and sound accountability system for the safe production of hazardous chemicals will be fully implemented, and the modernization of governance system and governance capacity on the workplace safety will be basically realized, the plan said.

The ministry announced last month to launch a year-long campaign to ensure the elimination of related risks during the relocation of the hazardous chemical industry in China.

The campaign seeks to strengthen the evaluation of new and existing projects and take action on those that fail to meet safety standards to ensure safety in the sector.

Practice and Training

Cardiopulmonary Resuscitation (CPR)

A normal CPR procedure uses chest compressions and ventilations (Fig. 4-3). The compressions push on the bone that is in the middle of the chest (sternum) and the ventilations are made pinching the victim's nose and blowing air mouth-to-mouth. If the victim is a baby, the rescuer would make the ventilations covering the baby's mouth and nose at the same time. It is recommended for all victims of any age a general compression-to-ventilation ratio of 30:2 (30 rhythmic compressions before each 2 ventilations).

As an exception for the normal compression-to-ventilation ratio of 30:2, if at least two trained rescuers are present, and the victim is a child, a ratio of 15:2 is preferred. And, according to the AHA 2015 Guidelines, the ratio in newborns is 30:2 if one rescuer is present and 15:2 if two rescuers are present. In an advanced airway treatment, such as an endotracheal tube or laryngeal mask airway, the artificial ventilation should occur without pauses in compressions, at a rate of 1 breath every 6 to 8 seconds (8-10 ventilations per minute).

In all the victims, the compression speed is of at least 100 compressions per minute. Recommended compression depth in adults and children is of 5 cm (2 inches), and in infants it is 4 cm (1.6 inches). In adults, rescuers should use two hands for the chest compressions (one on the top of the other), while in children one hand can be enough, and with babies the rescuer must use only two fingers.

There exist some plastic shields and respirators that can be used in the rescue breaths between the mouths of the rescuer and the victim, with the purposes of sealing a better vacuum and avoiding infections.

In some cases, the patient has experienced one of the failures in the rhythm of the heart (ventricular fibrillation and ventricular tachycardia) that can be corrected with the electric shock of a defibrillator. It is important then that someone asks for the defibrillator and to use it, which would be easy, because the common models of defibrillator (the AEDs) are automatic portable machines that guide to the user with recorded voice instructions along the process, and analyze the victim, and apply the correct shocks if they are needed. Besides, there exist written instructions of defibrillators that explain how to use them step-by-step.

The recommended order of normal cardiopulmonary resuscitation is the 'CAB': first 'Chest' (chest compressions), followed by 'Airway' (attempt to open the airway by performing a head tilt and a chin lift), and 'Breathing' (rescue breaths). Anyway, as of 2010, the Resuscitation Council (UK) was still recommending an 'ABC' order if the victim is a child. It can be difficult to determine the presence or absence of a pulse, so the pulse check has been removed for common providers and should not be performed for more than 10 seconds by healthcare providers.

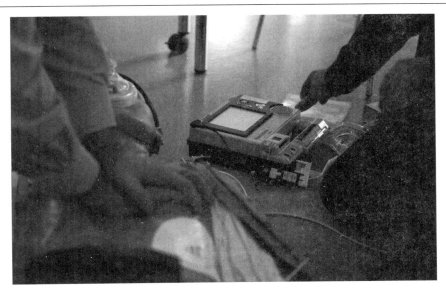

Fig. 4-3 CPR training: CPR is being administered while a second rescuer prepares for defibrillation.

Unit Five Chemical Engineering and Technology

化学工程技术

Lesson One   Primary Principles of Unit Operation
单元操作基本原理

Chemical engineering has to do with industrial processes in which raw materials are changed or separated into useful products. The chemical engineer must develop, design, and engineer both the complete process and the equipment used; choose the proper materials; operate the plants efficiently, safely, and economically; and see to it that products meet the requirements set by the customers. Chemical engineering is both an art and a science. Whenever science helps the engineer to solve a problem, science should be used. When, as is usually the case, science does not give a complete answer, it is necessary to use experience and judgment. The professional stature of an engineer depends on skill in utilizing all sources of information to reach practical solutions to processing problems.

The variety of processes and industries that call for the services of chemical engineers is enormous. Products of concern to chemical engineers range from commodity chemical such as sulfuric acid and chlorine to high-technology items such as high-strength composite materials, and genetically modified biochemical agents. The processes described in standard treatises on chemical technology and the process industries give a good idea of the field of chemical engineering, as does the 1988 report on the profession by the National Research Council.

Because of the variety and complexity of modern processes, it is not practicable to cover the entire subject matter of chemical engineering under a single head. The field is divided into convenient, but arbitrary, sectors. This text covers that portion of chemical engineering known as the unit operation.

An economical method of organizing much of the subject matter of chemical engineering is based on two facts: (1) although the number of individual processes is great, each one can be broke down into a series of steps, called operations, each of which in turn appears in process after process; (2) the individual operations have common techniques and are based on the same scientific principles. For example, in most processes solids and fluids must be moved; heat or other forms of energy must be transferred from one substance to another; and tasks such as drying, size reduction, distillation, and evaporation must be performed. The unit operation concept is this:

by studying systematically these operations themselves—operations that clearly cross industry and process lines—the treatment of all processes is unified and simplified.

The strictly chemical aspects of processing are studied in a companion area of chemical engineering called reaction kinetics. The unit operations are largely used to conduct the primarily physical steps of preparing the reactants, separating and purifying the products, recycling unconverted reactants, and controlling the energy transfer into or out of the chemical reactor.

The unit operations are as applicable to many physical processes as to chemical ones. For example, the process used to manufacture common salt consists of the following sequence of unit operations: transportation of solids and liquids, transfer of heat, evaporation, crystallization, drying, and screening. No chemical reaction appears in these steps. On the other hand, the cracking of petroleum, with or without the aid of catalyst, is a typical chemical reaction conducted on an enormous scale. Here the unit operations—transportation of fluids and solids, distillation, and various mechanical separations—are vital, and the cracking reaction could not be utilized without them. The chemical steps themselves are conducted by controlling the flow of material and energy to and form the reaction zone.

Because the unit operations are a branch of engineering, they are based on both science and experience. Theory and practice must combine to yield designs for equipment that can be fabricated, assembled, operated, and maintained. A balanced discussion of each operation requires that theory and equipment be considered together.

A number of scientific principles and techniques are basic to the treatment of the unit operations. Some are elementary physical and chemical laws such as conservation of mass and energy, physical equilibria, kinetics, and certain properties of matter.

The official international system of units is SI. Strong efforts are underway for its universal adoption as the exclusive system for all engineering and science, but old systems, particularly the centimeter-gram-second (cgs) and foot-pound-second (fps) engineering gravitational systems, are still in use and probably will be around for some time. The chemical engineer finds many physiochemical data given in cgs units; that many calculations are most conveniently done in fps units; and that SI units are increasingly encountered in science and engineering. Thus it becomes to the expert in the use of all the systems.

The original listing of the unit operation quoted above names twelve actions, not all of which are considered unit operations. Additional ones have been designated since then, at a modest rate over the years but recently at an accelerating rate. Fluid flow, heat transfer, distillation, humidification, gas absorption, sedimentation, classification, agitation, and centrifugation have been recognized. In recent increasing understanding of new techniques and adaptation of old but seldom used separative techniques has led to a continually increasing number of separations, processing operations, or steps in a manufacture that could be used without significant alteration in a variety of processes. This is the basis of a terminology of unit operation, which now offers us a list of techniques, all of which can not be covered in a reasonable text.

The typical chemical manufacturing operation involves a few chemical steps that are probably straightforward and well understood. Extensive equipment and operations are usually needed for refining or further preparing the often complex mixture for use as an end product. This result is that the work of the typical process engineer is much more concerned with physical changes than with chemical reaction. The importance of the chemical reactions must not be overlooked because of the economic importance of small improvements in percentage yield from chemical reaction. In many cases a relatively small percentage improvement in yield may economically considerably more extensive processing operations and equipment.

All unit operations are base on principles of science that are translated into industrial applications in various fields of engineering. The flow of fluids, for instance, has been studied extensively in theory under the name of hydrodynamics or fluid mechanics. It has been an important part of the work of civil engineers under the name of hydraulics and is of major importance in sanitary engineering. Problems of water supply and control has been met by every civilization.

Throughout industry, one finds examples of most of the unit operations in applications that are in the province of other engineering fields. The chemical engineer must carry out many unit operations on materials of widely varying physical and chemical properties under extremes of conditions such as temperature and pressure. The unit operations used to separate mixtures into more or less pure substances is unique to chemical engineering. The materials being processed may be naturally occurring mixtures or they may be the products of chemical reactions, which virtually never yield a pure substance.

Words and Expressions

equipment [ɪˈkwɪpmənt]   *n.* 设备
economically [iːkəˈnɒmɪkəlɪ]   *adv.* 经济的
utilize [juːˈtɪlaɪz]   *v.* 利用
chlorine [ˈklɔːriːn]   *n.* 氯
arbitrary [ˈɑːbɪtrərɪ]   *adj.* 任意的
polymeric [ˌpɒlɪˈmerɪk]   *adj.* 聚合的，聚合物的
electronics [ɪlekˈtrɒnɪks]   *n.* 电子学（单数）
composite [ˈkɒmpəzɪt, ˈkɒmpəzaɪt]   *adj.* 合成的，复合的；*n.* 合成物，复合材料；*vt.* 合成
break down   打破
a series of   一系列
solid [ˈsɒlɪd]   *n.* 固体
liquid [ˈlɪkwɪd]   *n.* 液体
drying [ˈdraɪɪŋ]   *n.* 干燥
size reduction   粉碎
distillation [ˌdɪstɪˈleɪʃən]   *n.* 蒸馏
evaporation [ɪˌvæpəˈreɪʃən]   *n.* 蒸发
reactant [riːˈæktənt]   *n.* 反应物
petroleum [pɪˈtrəʊlɪəm]   *n.* 石油

Notes

1. unit operation　单元操作
2. chemical engineering　化学工程
3. raw material　原材料
4. biochemical agents　生化试剂
5. chemical technology　化学工艺
6. commodity chemical　日用化工
7. sulfuric acid　硫酸
8. chemical reactor　化学反应器
9. physical process　物理过程
10. chemical process　化学过程
11. common salt　食盐
12. reaction kinetics　反应动力学

13. The chemical engineer must develop, design, and engineer both the complete process and the equipment used; choose the proper materials; operate the plants efficiently, safely, and economically; and see to it that products meet the requirements set by the customers.

译文：化学工程师必须开发、设计及制造完整的过程和设备；选择适当的材料；简单地、安全地、经济地进行工厂操作；保证产品符合规定的要求。

14. Products of concern to chemical engineers range from commodity chemical such as sulfuric acid and chlorine to high-technology items such as high-strength composite materials, and genetically modified biochemical agents.

译文：化学工程师关注的产品范围从商用化工产品如硫酸和氯化物到高科技项目，如高强度的复合材料和转基因生物制剂。

15. An economical method of organizing much of the subject matter of chemical engineering is based on two facts: (1) although the number of individual processes is great, each one can be broke down into a series of steps, called operations, each of which in turn appears in process after process; (2) the individual operations have common techniques and are based on the same scientific principles.

译文：成功组织化学工程项目包括 2 个部分:(1)每个独立过程都可以分解成一系列步骤，称为单元操作,每一步骤依次出现；(2)每个独立的操作都具有相同的技术，并且基于相同的科学原理。

16. For example, in most processes solids and fluids must be moved; heat or other forms of energy must be transferred from one substance to another; and tasks such as drying, size reduction, distillation, and evaporation must be performed.

译文：比如说，在许多过程中，固体和液体必须被输送；热量或者其他形式的能量必须从一个物质传递至另一个物质；必须实施干燥、粉碎、蒸馏、蒸发等操作。

17. The process used to manufacture common salt consists of the following sequence of unit operations: transportation of solids and liquids, transfer of heat, evaporation, crystallization, drying, and screening.

译文：盐的制造过程包含这样一些单元操作：固体和液体的输送，热量传递，蒸发，结晶，干燥和筛分。

Reading Comprehension

1. How to organize any subjects of chemical engineering?
2. What are the unit operations based on?
3. What are the examples of the unit operations?
4. What is the unit operations used to do?

Reading Material

New Technologies in Unit Operations 单元操作新技术

While technical advances and efficiency improvements in specific unit operation are occurring all the time, the big story is the hybridization of processes. Combining individual unit operations, such as reaction, separation, and heat exchange, into larger, concurrent operations will be major trend in upcoming years. Technologies such as reactive distillation, catalytic membranes, and phase-transfer catalysis all represent examples of hybridized operations is a huge reduction in capital expenditure—typically to one-tenth to one-fifth of the investment for a traditional setup. In practice, at least at first while confidence in the performance and reliability of the combined operations builds, companies usually will run the newer, hybridized operations in parallel with older processes for the same produce. The typical unit operations are shown in Fig.5-1.

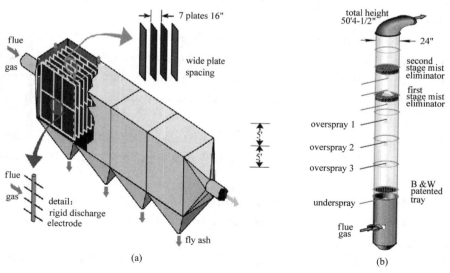

Fig.5-1 Typical unit operations

In addition to their significant reduction in capital outlay, combined reaction/separation processes offer two other major advantages: reduction of unwanted reaction byproducts, and improvement of yields for reactions with low equilibrium constants.

With environmental issues driving so many of the changes on the CPI, technologies that help reduce reaction byproducts clearly are of increasing value. With combined reaction/separation, highly reactive feedstock does not have a chance to react with products because the products are separated out immediately after formation. This also reduces waste of feedstock and product. Immediate removal of reaction products also will enable low equilibrium constant reactions that are not commercially feasible under normal conditions to proceed much further. Indeed, reactions that go to only a few percent conversions can be forced to 100% conversion through combined reaction/separation. Several processes in this category are currently under development, and large commercial operations are currently being designed and built for them.

A couple of examples of recent applications of these hybrid processes illustrate the benefits of this new approach to unit operations. Reactive distillation is proving very useful for the production of ethers, such as methyl *tert*-butyl ether (MTBE), *tert*-amyl methyl ether (TAME), and ethyl *tert*-butyl (ETBE), which increasingly are employed to boost oxygenate content in gasoline. Here, the preferred temperature range for the catalyst is the same as that for the distillation of ethers from reactants and inerts. Thus, reactive distillation provides energy savings, as well.

Catalytic membranes are a newer example of combined reaction/separation processes. In this scheme, the catalyst material also acts as a sieving system to separate reaction products as they are formed. A key advantage of membrane separation processes is their energy efficiency; they also particularly suit heat-sensitive material, such as pharmaceuticals and foodstuffs. Dense catalytic membranes, which separate, for example, on the basis of gas diffusivity, are closer to commercial deployment than are porous membranes, which separate on the basis of molecular size. Robust processes have not yet emerged, however, for synthesizing defect-free porous membranes with appropriate composition and pore size range for the variety of catalytic reactions and gas adsorption applications that exist in the CPI.

A second and newer trend in unit operations is the advent of the minireactor. The idea of "desktop chemical manufacturing" takes the large scale, continuous commodity plant and scales it down for the manufacture of specialty chemicals such as pharmaceuticals. Instead of batch processing different chemicals one after another in the same equipment, a plant might consist of several small systems running continuously throughout the year. The advantages include better consistency of product, simpler scheduling and monitoring, and standardized small-scale equipment. Although a consensus has not yet emerged about the overall economic value of the minireactor approach, early industrial trials have shown that there may be some significant advantages.

This minireactor approach has even been extended to the idea of the microreactor. University and industrial researchers have developed a prototype microfluidic reactor that measures no more than 2 square cm. Proposed reaction systems for the unit include hydrogen and methane oxidation, ethylene epoxidation, and phosgene synthesis. Once optimized, such a microreactor then can be scaled up to commercial proportions through simple replication and arrangement of the individual units. The potential advantages of such a system include better process safely relative to macroscale reactors, and improved ability to integrate control, sensor, and reactor functionality.

Words and Expressions

hybridization [ˌhaɪbrɪdaɪˈzeɪʃən]	n. 杂交，杂化
concurrent [kənˈkʌrənt]	adj. 并流的，顺流的
byproduct [ˈbaɪˌprɒdʌkt]	n. 副产物
feedstock [ˈfiːdstɒk]	n. 原料
product [ˈprɒdəkt]	n. 产品
conversion [kənˈvɜːʃən]	n. 转化，转换
ether [ˈiːθə]	n. 醚
inert [ɪˈnɜːt]	n. 惰性组分
sieving [sɪvɪŋ]	n. 筛分
diffusivity [dɪfjʊˈsɪvɪtɪ]	n. 扩散性，扩散系数
deployment [dɪˈplɔɪmənt]	n. 使用，利用，推广应用
robust [rəˈbʌst]	adj. 加强的，增强的
adsorption [ædˈsɔːpʃən]	n. 吸附
consensus [kənˈsensəs]	n. 一致
trial [ˈtraɪəl]	n. 试验，审判
approach [əˈprəʊtʃ]	n. 途径，方法，手段
replication [ˌreplɪˈkeɪʃən]	n. 重复试验

Lesson Two Fluid Flow Phenomena
流体流动现象

The behavior of fluids is important to process engineering generally and constitutes one of the foundations for the study of unit operations. An understanding of fluids is essential, not only for accurately treating problems in the movement of fluids through pipes, pumps, and all kinds of process equipment but also for the study of heat flow and the many separation operations that depend on diffusion and mass transfer.

The branch of engineering science that has to do with the behavior of fluids—and fluids are understood to include liquids, gases, and vapors—is called fluid mechanics. Fluid mechanics in turn is part of a larger discipline called continuum mechanics, which also includes the study of stressed solids.

The behavior of a flowing fluid depends strongly on whether the fluid is under the influence of solid boundaries. In the region where the influence of the wall is small, the shear stress may be negligible and the fluid behavior may approach that of an ideal fluid, one that is incompressible and has zero viscosity. The flow of such an ideal fluid is called potential flow and is completely described by the principles of Newtonian mechanics and conservation of mass. The mathematical theory of potential flow is highly developed but is outside the scope of this book. Potential flow has two important characteristics: (1) neither circulations nor eddies can form within the stream, so that potential flow is also called irrotational flow; and (2) friction can not develop, so that there is no dissipation of mechanical energy into heat.

Potential flow can exist at distances not far from a solid boundary. A fundamental principle of fluid mechanics, originally stated by Prandtl in 1904, is that, except for fluids moving at low velocities or possessing high viscosities, the effect of the solid boundary on the flow is confined to a layer of the fluid immediately adjacent to the solid wall. This layer is called the boundary layer, and shear and shear forces are confined to this part of the fluid. Outside the boundary layer, potential flow survives. Most technical flow processes are best studied by considering the fluid stream as two parts, the boundary layer and the remaining fluid. In some situations such as flow in a converging nozzle, the boundary layer may be neglected; and in others, such as flow through pipes, the boundary layer fills the entire channel, and there is no potential flow.

Within the current of an incompressible fluid under the influence of solid boundaries, four important effects appear: (1) the coupling of velocity-gradient and shear-stress fields, (2) the onset of turbulence, (3) the formation and growth of boundary layers, and (4) the separation of boundary layers from contact with the solid boundary.

In the flow of compressible fluids pass solid boundaries, additional effects appear, arising from the significant density changes that are characteristic of compressible fluids.

When a stream of fluid is flowing in bulk past a solid wall, the fluid adheres to the solid at the actual interface between solid and fluid. The adhesion is a result of the force fields at the boundary, which are also responsible for the interfacial tension between solid and fluid. If, therefore, the wall is at rest in the reference frame chosen for the solid-fluid system, the velocity of the fluid at the interface is zero. Since at distances away from the solid the velocity is not zero, there must be variations in velocity from point to point in the flowing stream. Therefore, the velocity at any point is a function of the space coordinates of that point, and a velocity field exists in the space occupied by the fluid. The velocity at a given location may also vary with time. When the velocity at each location is constant, the field is invariant with time and the flow is said to be steady.

Velocity is a vector, and in general, the velocity at a point has three components, one for each space coordinate. In many simple situations all velocity in the field are parallel or practically so and only one velocity component, which may be taken as a scalar, is required. This situation, which obviously is much simpler than the general vector field, is called one-dimensional flow; an example is steady flow through straight pipe.

Words and Expressions

fluid ['fluːɪd]   *n.* 流体
phenomena [fɪ'nɒmɪnə]   *n.* 现象（phenomenon 的复数）
viscosity [vɪs'kɒsɪtɪ]   *n.* 黏度
circulation [ˌsɜːkjʊ'leɪʃən]   *n.* 循环
eddy ['edɪ]   *n.* 漩涡
irrotational flow   无旋流

dissipation [ˌdɪsɪ'peɪʃən]   n. 损耗
nozzle ['nɒzl]   n. 喷嘴，排气口
neglect [nɪ'glekt]   vt. 忽视
turbulence ['tɜːbjʊləns]   n. 湍流
adhesion [əd'hiːʒən]   n. 附着
velocity [vɪ'lɒsɪtɪ]   n. 速度
assumption [ə'sʌmpʃən]   n. 假定

Notes

1. fluid flow 流体流动
2. potential flow 潜流
3. ideal fluid 理想流体
4. incompressible fluid 不可压缩流体
5. boundary layer 边界层
6. interfacial tension 界面张力
7. space coordinate 空间坐标
8. one-dimensional flow 一维流动

9. The behavior of a flowing fluid depends strongly on whether the fluid is under the influence of solid boundaries. In the region where the influence of the wall is small, the shear stress may be negligible and the fluid behavior may approach that of an ideal fluid, one that is incompressible and has zero viscosity.

译文：流体的流动取决于流体是否处于固体边界的影响下。在这个地区壁的影响很小，可以忽略剪应力，此时流体的流动类似于具有不可压缩性和黏度为零的理想流体。

10. Potential flow has two important characteristics: (1) neither circulations nor eddies can form within the stream, so that potential flow is also called irrotational flow; and (2) friction can not develop, so that there is no dissipation of mechanical energy into heat.

译文：潜流具有两个重要的特点：(1)与水流既不形成流通也不生成漩涡，所以潜流也可称为无旋流；(2)不能形成摩擦，因此没有耗散的机械能转化为热能。

11. A fundamental principle of fluid mechanics, originally stated by Prandtl in 1904, is that, except for fluids moving at low velocities or possessing high viscosities, the effect of the solid boundary on the flow is confined to a layer of the fluid immediately adjacent to the solid wall.

译文：流体力学的基本原则，最初是由 Prandtl 在 1904 年提出的，除了流体移动速度较低或高黏度的影响，流体在固体边界层上的影响限于临近其的一层流体。

12. Within the current of an incompressible fluid under the influence of solid boundaries, four important effects appear: (1) the coupling of velocity-gradient and shear-stress fields, (2) the onset of turbulence, (3) the formation and growth of boundary layers, and (4) the separation of boundary layers from contact with the solid boundary.

译文：受压流体在固体边界的影响下，有四个重要作用：(1)速度梯度和剪应力区域的耦合；(2)湍流的开始；(3)边界层的形成与发展；(4)将边界层与固体边壁相连处分离。

13. In many simple situations all velocity in the field are parallel or practically so and only one

velocity component, which may be taken as a scalar, is required.

译文：许多简单的情况下，在该领域所有的速度是平行的或几乎平行的，仅仅只需要一个被看成是标量的速度分量。

Reading Comprehension

1. What do the behavior of a flowing fluid depend on?
2. What are the characteristics of the potential flow?
3. What is the fundamental principle of the fluid mechanics?
4. Within the current of an incompressible fluid under the influence of solid boundaries, what do the four important effects appear?

Reading Material

Turbulent Flow 湍流

There are several situations that can lead to turbulent flow. One situation just described is that of rapid flow of a fluid past a solid surface. This situation leads to unstable, self-amplifying velocity fluctuations, which form in the fluid in the vicinity of the wall and spread outward into the main fluid stream. In a similar manner, turbulent eddies are formed from the velocity gradients established between a fast-moving fluid and a slower-moving fluid. A third general way in which turbulence is induced is by the relative movement of an object through the fluid streams. Examples of this last case are an impeller blade on an agitator and a sphere or cylinder falling through a fluid medium. These situations cause eddies to form in the wake, resulting in an increase in the resistance of the movement of an object.

In the case of stirred vessels, turbulence can be very intense near the tips of the rotor blades. Most of the turbulence throughout the vessel arises from velocity gradients, whereby portions of high-velocity fluid are thrown from impeller blades into slowermoving fluid. Some of this turbulence is attributed, however, to the high shearing over the blades themselves, which creates separation behind each blade or neighboring baffles.

Another mixing-related turbulence arises from submerged jets. In submerged jets, the fluid is expelled from a nozzle into a mass of miscible fluid, which is essentially at rest with respect to the discharging stream. The free jet expands outward in the form of a cone, entraining large amounts of the surrounding fluid at the periphery of the jet. The sizes of eddies formed are relatively uniform across any given section of such a jet. When a submerged jet is introduced to an immiscible fluid, the stream is referred to as a restrained turbulent jet. Turbulent eddies protrude from the sides of the free jet but are restrained from breaking away by surface tension forces. This type of turbulence decays quickly, as fluctuations are damped by the elastic forces associated with surface tension effects.

It should be noted that the turbulent condition set by $N_{Re} > 4000$ is a general rule and one based mainly on experience with pipe flow. In fact, turbulent conditions can be generated at much lower Reynolds numbers. One example is flow in a corrugated conduit, where vortex shedding at the angularities can induce strong eddying at Reynolds numbers of only a few hundred. This principle is also applied to inline static mixers used for mixing two liquid streams or a chemical additive and a liquid. By means of fins, baffles or an arrangement of packing material through which fluids flow vortex shedding occurs, promoting eddies assisting in the necessary mass transfer. Another example of this application is in heat exchangers used for viscous liquids. Here, only moderate flowrate needed to promote eddy formation in narrow tubes. This promotes good heat transfer even through Reynolds numbers are only as high as perhaps 500-600. In these applications, the sharpness of the angularities is important since eddies shed much more readily from sharp corners than from smooth surfaces.

Words and Expressions

turbulent [ˈtɜːbjʊlənt]   adj. 狂暴的，湍急的
eddy [ˈedɪ]   n. 逆流，涡流
shedding [ˈʃedɪŋ]   n. 脱落，流出，散发
additive [ˈædɪtɪv]   n. 添加剂
fin [fɪn]   n. 翅片
velocity [vɪˈlɒsɪtɪ]   n. 速度，速率
fluctuation [ˌflʌktjʊˈeɪʃən]   n. 波动，起伏
impeller [ɪmˈpelə]   n. 推进者，轮叶
blade [bleɪd]   n. 刀锋，刀口
gradient [ˈɡreɪdɪənt]   n. 梯度

Lesson Three Chemical Engineering
化学工程

Chemical engineering is the development of processes and the design and operation of plants in which materials undergo change in physical or chemical state on a technical scale. Applied throughout the process industries, it is founded on the principles of chemistry, physics, and mathematics. The laws of physical chemistry and physics govern the practicability and efficiency of chemical engineering operations. Energy changes, deriving from thermodynamic considerations, are particularly important. Mathematics is a basic tool in optimization and modeling. Optimization means arranging materials, facilities, and energy to yield as productive and economical an operation as possible. Modeling is the construction of theoretical mathematical prototypes of complex process systems, commonly with the aid of computers.

Chemical engineering is as old as the process industries. Its heritage dated from the fermentation and evaporation processes operated by early civilizations. Modern chemical engineering emerged with the development of large-scale, chemical-manufacturing operations in the second half of the 19th century. Throughout its development as an independent discipline, chemical engineering has been directed toward solving problems of designing and operating large plants for continuous production.

Manufacture of chemicals in the mid-19th century consisted of modest craft operations. Increase in demand, public concern at the emission of noxious effluents, and competition between rival processes provided the incentives for greater efficiency. This led to the emergence of combines with resources for larger operations and caused the transition from a craft to a science-based industry. The result was a demand for chemists with knowledge of manufacturing processes, known as industrial chemists or chemical technologists. The term chemical engineer was in general use by about 1900. Despite its emergence in traditional chemicals manufacturing, it was through its role in the development of the petroleum industry that chemical engineering became firmly established as a unique discipline. The demand for plants capable of operating physical separation processes continuously at high levels of efficiency was a challenge that could not be met by the traditional chemist or mechanical engineer.

A landmark in the development of chemical engineering was the publication in 1901 of the first textbook on the subject, by Gorge E. Davis, a British chemical consultant. This concentrated on the design of plant items for specific operation. The notion of a processing plant encompassing a number of operations, such as mixing, evaporation, and filtration, and of these operations being essentially similar, whatever the product, led to the concept of unit operations. This was first enunciated by the American chemical engineer Arthur D. Little in 1915 and formed the basis for a classification of chemical engineering that dominated the subject for the next 40 years. The number of unit operation — the building blocks of a chemical plant — is not large. The complexity arises from the variety of conditions under which the unit operations are conducted. In the same way that a complex plant can be divided into basic unit operations, so chemical reactions involved in the processes (e.g. polymerizations, esterifications, and nitrations), having common characteristics. This classification into unit processes brought rationalization to the study of process engineering.

The unit approach suffered from the disadvantage inherent in such classifications: a restricted outlook based on existing practice. Since World War Two, closer examination of the fundamental phenomena involved in the various unit operations has shown these to depend on the basic laws of mass transfer, heat transfer, and fluid flow. This has given unity to the diverse unit operations and has led to the development of chemical engineering science in its own right; as a result, many applications have been found in fields outside the traditional chemical industry.

Study of the fundamental phenomena upon which chemical engineering is based has necessitated their description in mathematical form and has led to more sophisticated mathematical techniques. The advent of digital computers has allowed laborious design calculations to be performed rapidly, opening the way to accurate optimization of industrial processes. Variations due to different parameters, such as energy source used, plant layout, and environmental factors, can be predicted accurately and quickly so that the best combination can be chosen.

Chemical Engineering Functions

Chemical engineers are employed in the design and development of both processes and plant items. In each case, data and predictions often have to be obtained or confirmed with pilot experiments. Plant operation and control is increasingly the sphere of the chemical engineer rather than the chemist. Chemical engineering provides an ideal background for the economic evaluation of new projects and, in the plant construction sector, for marketing.

Branches of Chemical Engineering

The fundamental principles of chemical engineering underlie the operation of processes extending well beyond the boundaries of the chemical industry, and chemical engineers are employed in a range of operations outside traditional areas. Plastics, polymers, and synthetic fibers involve chemical reaction engineering problems in their manufacture, with fluid flow and heat transfer considerations dominating their fabrication. The dyeing of a fiber is a mass-transfer problem. Pulp and paper manufactures involve considerations of fluid flow and heat transfer. While the scale and materials are different, these again are found in modern continuous production of foodstuffs. The pharmaceuticals industry presents chemical engineering problems, the solutions of which have been essential to the availability of modern drugs. The nuclear industry makes similar demands on the chemical engineer, particularly for fuel manufacture and reprocessing. Chemical engineers are involved in many sectors of the metals processing industry, which extends from steel manufacture to separation of rare metals.

Further applications of chemical engineering are found in the fuel industries. In the second half of the 20th century, considerable numbers of chemical engineers have been involved in space exploration, from the design of fuel cells to the manufacture of propellants. Looking to the future, it is probable that chemical engineering will provide the solution to at least two of the world's major problems: supply of adequate fresh water in all regions through desalination of seawater and environmental control through prevention of pollution.

Words and Expressions

Chemical Engineering	化学工程
theoretical mathematical prototypes	理论数学模型
thermodynamics [ˈθɜːməʊdaɪˈnæmɪks]	n. 热力学
heritage [ˈherɪtɪdʒ]	n. 继承物，遗产
craft [krɑːft]	n. 手艺，技艺
rationalization [ˌræʃənəlaɪˈzeɪʃən]	n. 合理化
foodstuff [ˈfuːdstʌf]	n. 食品
desalination [diːˌsælɪˈneɪʃən]	n. 脱盐
fuel cells	燃料电池

Notes

1. In the same way that a complex plant can be divided into basic unit operations, so chemical reactions involved in the processes (e.g. polymerizations, esterifications, and nitrations), having

common characteristics.

译文：与复杂的工厂可划分为基本的单元操作一样，过程工业中涉及的化学反应也可以分成一定的单元过程（如聚合、酯化和硝化），它们具有共同的特征。

2. Plastics, polymers, and synthetic fibers involve chemical reaction engineering problems in their manufacture, with fluid flow and heat transfer considerations dominating their fabrication.

译文：塑料、聚合物和合成纤维在生产中涉及化学反应工程问题，其中流体流动和传热是生产中主要考虑的因素。

3. In the second half of the 20th century, considerable numbers of chemical engineers have been involved in space exploration, from the design of fuel cells to the manufacture of propellants.

译文：20世纪下半叶，从燃料电池的设计到推进剂的生产，相当数量的化学工程师参与了空间的探索。

4. Further applications of chemical engineering are found in the fuel industries.

译文：化学工程的深入应用是在燃料工业。

Reading Comprehension

1. Why is chemical engineering still focused on nowadays?
2. What are the basic laws of chemical engineering science?
3. What functions and branches of chemical engineering do you know?
4. Do you think that the development of chemical engineering will lead to environment pollution?

Reading Material

What is Chemical Engineering? 什么是化学工程

Society can associate civil engineers with huge building complexes and bridges, electronic and electrical engineers with telecommunications and power generation, and mechanical engineers with advanced machinery and automobiles. However, chemical engineers have no obvious monuments which create an immediate awareness of discipline in the public mind. Nevertheless, the range of products in daily use which are efficiently produced as a result of the application of chemical engineering expertise is enormous. The list given in Table 5-1 is not exhaustive, and any reader who grasps the key element, which involves the conversion of raw materials into a useful product, will be able to extend it. Although the products are unglamorous, the creation and operation of cost-effective processes to produce them is often challenging and exciting.

The term "chemical engineering" implies that the person is primarily an engineer whose first professional concern is with manufacturing processes — making something, or making some process work. The adjective "chemical" implies a particular interest in process which involves chemical changes. While the main term is correct, the adjective is too restrictive and the literal definition will not suffice. Taken at face value, it would exclude many areas in which chemical engineers have made their mark, for example, textiles, nuclear fuels and the food industry. Thus the Institution of Chemical Engineers defines chemical engineering as "that branch of engineering which is concerned with processes in which materials undergo a required change in composition, energy content or physical state: with the means of processing; with the resulting products, and with their application to useful ends". It is perhaps too presumptuous to insist that the term "process engineer" should replace the term "chemical engineer", and so the two will be used synonymously.

It should also be noted that large-scale processes involving biological systems (such as waste water treatment and production of protein) fit the definition as well as traditional chemical processes such as the production of fertilizers and pharmaceuticals. The work of chemical engineers will be examined by the way of four case studies in the second part of this chapter, but to complete the definition, explicit mention of the concern that process operations be both safe and economic must be made.

A jocular, helpful, but very incomplete description is that, "a chemical engineer is a chemist who is aware of money". Although this neglects many, if not most, aspects of a chemical engineer's training, it does illustrate one important facet of any engineer's work. When working on a large scale, the cost of equipment and raw materials are more important than the cost of manpower. While the research chemist might use aqueous potassium hydroxide to neutralize acids, because it is pure and readily available, the chemical engineer will specify a cheaper alternative, provided that it serves the same purpose. Two obvious substitutes are aqueous sodium hydroxide, which is available at less than a tenth of the cost, or calcium hydroxide, which is even cheaper, but harder to handle. In choosing between these hydroxide is sparingly soluble against the higher price of sodium hydroxide.

Table 5-1 A selection of everyday products whose manufacture involves
The application of chemical engineering

Product grouping or production
1. Household products in daily use
2. Health care products
3. Automotive fuels / Petroleum refining
4. Other chemicals in daily use
5. Metals
6. Electronics
7. Fats and oils
8. Daily products
9. Gas treatment and transmission

Some of the more familiar examples
1. Clothes, curtains, sheets, blankets
2. Fertilizers, insecticides
3. Steel manufacture, Zinc production
4. Beer, certain antibiotics such as penicillin, yogurts
5. Milk, butter, cheese, baby food
6. Salad and cooking oils, margarine, soap
7. Raw materials, silicon, dopants

Lesson Four The Anatomy of a Chemical Manufacturing Process
化工生产过程分解

The basic components of a typical chemical process are shown in Fig. 5-2, in which each block represents a stage in the overall process for producing a product from the raw materials. Fig.5-2 represents a generalized process; not all the stages will be needed for any particular process and the complexity of each stage will depend on the nature of the process. Chemical engineering design is concerned with the selection and arrangement of the stages, and the selection, specification and design of the equipment required to perform the stage functions.

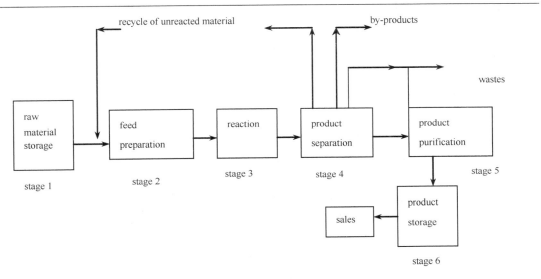

Fig.5-2 A generalized process

Stage 1. Raw material storage

Unless the raw material (also called essential materials, or feedstocks) are supplied as intermediate products (intermediates) from a neighboring plant, some provision will have to be made to hold several days' or weeks' storage to smooth out fluctuations and interruptions in supply. Even when the material come from an adjacent plant, some provision is usually made to hold a few hours, or even a few days, supplying to decouple the processes. The storage required will depend on the nature of the raw materials, the method of delivery, and what assurance can be placed on the continuity of supply. If materials are delivered by ship (tanker or bulk carrier), several weeks' stock may be necessary; if they are received by road or rail, in smaller lots, less storage will be needed.

Stage 2. Feed preparation

Some purification, and preparation, of raw material will usually be necessary before they are sufficiently pure, or in the right form, to be fed to the reaction stage. For example, acetylene generated by the carbide process contains arsenical and sulphur compounds, and other impurities, which must be removed by scrubbing with concentrated sulphuric acid (or other processes) before it is sufficiently pure for reaction with hydrochloric acid to produce dichloroethane. Liquid feeds will need to be vaporized before being fed to gas-phase reactors and solids may need crushing, grinding and screening.

Stage 3. Reactor

The reaction stage is the heart of a chemical manufacturing process. In the reactor the raw materials are brought together under conditions that promote the production of the desired product; invariably, by-products and unwanted compounds (impurities) will also be formed.

Stage 4. Product separation

In this first stage alter the reactor the products and by-products are separated from any unreacted material. If in sufficient quantity, the unreacted material will be recycled to the reactor. They may be returned directly to the reactor or to the feed purification and preparation stage. The by-products may also be separated from the products at this stage.

Stage 5. Purification

Before sale the main product will usually need purification to meet the product specification. If produced in economic quantities, the by-products may also be purified for sale.

Stage 6. Product storage

Some inventory of finished product must be held to match production with sales. Provision for product packaging and transport will also be needed, depending on the nature of the product. Liquids will normally be dispatched in drums and in bulk tankers (road, rail and sea), solids in sacks, cartons or bales.

The stack held will depend on the nature of the product and the market.

1. Ancillary Processes

In addition to the main process stages shown in Fig.5-2, provision will have to made for the supply of the services (utilities) needed, such as process water, cooling water, compressed air, steam. Facilities will also be needed for maintenance, firefighting, offices and other accommodation, and laboratories.

2. Continuous and Batch Processes

Continuous processes are designed to operate 24 hours a day, 7 days a week, through-out the year. Some down time will be allowed for maintenance and, for some processes, catalyst regeneration. The plant attainment, that is, the percentage of the available hours in a year that the plant operates, will usually be 90% to 95%.

$$\text{Attainment\%} = \frac{\text{hours operated}}{24 \times 365}$$

Batch processes are designed to operate intermittently. Some or all the process units are frequently shut down and started up.

Continuous processes will usually be more economical for large scale production. Batch processes are used where some flexibility is wanted in production rate or product specification.

The choice between batch and continuous operation will not be clear cut, but the following rules can be used as a guide.

Continuous

1. Production rate greater than 5×100kg /h
2. Single product
3. No severs fouling
4. Good catalyst life
5. Proven processes design
6. Established market

Batch

1. Production rate less than 5×100kg/h
2. A range of products or product specifications
3. Severe fouling
4. Short catalyst life
5. New product
6. Uncertain design

Words and Expressions

anatomy [əˈnætəmɪ]	n. 分解
component [kəmˈpəʊnənt]	n. 组成
a typical chemical process	典型化工生产工程
essential materials	原料
gas-phase reactor	气相反应器
ancillary [ænˈsɪlərɪ]	n. 辅助
attainment [əˈteɪnmənt]	n. 开车率，开工率
firefighting [ˈfaɪəfaɪtɪŋ]	n. 消防
by-products	副产品
hydrochloric acid	盐酸
batch processes	间隙生产过程
continuous processes	连续生产过程
catalyst life	催化剂寿命

Notes

1. Chemical engineering design is concerned with the selection and arrangement of the stages, and the selection, specification and design of the equipment required to perform the stage functions.

译文：化学工程设计是关于这些步骤的选择和组合，以及实现这些步骤功能的设备选型、技术参数和设计。

2. If in sufficient quantity, the unreacted material will be recycled to the reactor.

译文：如果数量足够多，未反应的原料将返回反应器循环使用。

3. Before sale the main product will usually need purification to meet the product specification.

译文：在销售之前，主产品通常需要精制以达到产品的技术指标。

4. If produced in economic quantities, the by-products may also be purified for sale.

译文：如果考虑经济数量，副产品也要提纯销售。

Reading Comprehension

1. How many stages are there in typical chemical manufacturing process?
2. What kinds of condition are suited to using batch and continuous processes?
3. Do you think that the ancillary process is necessary in a special chemical manufacturing process?

Reading Material

Chemical Process Safety 化工过程安全

In 1987, Robert M. Solow, an economist at the Massachusetts Institute of Technology, received the Nobel Prize in economics for his work in determining the sources of economic growth. Professor Solow concluded that the bulk of an economy's growth is the result of technological advances.

It is reasonable to conclude that the growth of an industry is also dependent on technological advances. This is especially true in the chemical industry, which is entering an era of more complex processes: higher pressure, more reactive chemicals, and exotic chemistry.

More complex processes require more complex safety technology. Many industrialists even believe that the development and application of safety technology is actually a constraint on the growth of the chemical industry. As chemical process technology becomes more complex, chemical engineers will need a more detailed and fundamental understanding of safety. H. H. Fawcett has said that to know is to survive and to ignore fundamentals is to court disaster.

Since 1950, significant technological advances have been made in chemical process safety. Today, safety is equal in importance to production and has developed into a scientific discipline which includes many highly technical and complex theories and practices. Examples of the technology of safety include:

a) Hydrodynamic models representing two-phase flow through a vessel relief.

b) Dispersion models representing the spread of toxic vapor through a plant after a release.

c) Mathematical techniques to determine the various ways that processes can fail, and the probability of failure.

Recent advances in chemical plant safety emphasize the use of appropriate technological tools to provide information for making safety decisions with respect to plant design and operation. The word safety used to mean the older strategy of accident prevention through the use of hard hats, safety shoes, and a variety of rules and regulations. The main emphasis was on worker safety. Much more recently, safety has been replaced by loss prevention. This term includes hazard identification, technical evaluation, and the design of new engineering features to prevent loss. The word safety and loss prevention will be used synonymously throughout for convenience. Safety, hazard, and risk are frequently-used terms in chemical process safety. Their definitions are

a) Safety or loss prevention is the prevention of accidents by the use of appropriate technologies to identify the hazards of a chemical plant and to eliminate them before an accident occurs.

b) A hazard is anything with the potential for producing an accident.

c) Risk is the probability of a hazard resulting in an accident.

Chemical plants contain a large variety of hazards. First, there are the usual mechanical hazards that cause worker injuries from tripping, failing, or moving equipment. Second, there are chemical hazards. These include fire and explosion hazards, reactivity hazards, and toxic hazards.

As will be shown later, chemical plants are the safest of all manufacturing facilities. However, the potential always exists for an accident of catastrophic proportions. Despite substantial safety programs by the chemical industry, headlines of the type shown in Figure continue to appear in the newspapers.

A successful safety program requires several ingredients. These ingredients are

a) Safety knowledge

b) Safety experience

c) Technical competence

d) Safety management support

e) Commitment

Lesson Five   Catalysts for Industrial Processes
工业催化剂

The criteria for an industrially successful catalyst are very stringent. The two most important considerations are activity and durability; suitability for basic study is not a factor, and this has retarded progress in our understanding of how such technical catalysts really operate. Let us consider future the two prime requirements.

First, the catalyst must be able to effect the desired reaction at an acceptable rate under conditions of temperature and pressure that are practicable. Chemical technology has advanced to the point where temperatures as high as 1,600K and pressure up to 350 atm (35MPa) present no difficulties and thermodynamic considerations sometimes necessitate their use to attain reasonable equilibrium yields of products. If however good yields can be obtained at low temperatures and pressures, then there is every incentive to find a catalyst that will operate under the mildest possible conditions, since the use of extreme conditions is very costly. It is concurrently important that side-reactions are minimal, especially those leading to poisoning or deactivation through carbon deposition.

Second, the catalyst must be able to sustain the desired reaction over prolonged periods: in some processes, a catalyst life of several years is not uncommon, and is economically necessary. Clearly the longer it lasts, the smaller will be the contribution that its initial cost makes to the overall cost of the process. Initial cost is rarely of over-riding importance: it is usually cheaper in the long run to use an expensive catalyst that will last a long time than a cheap one that has to be replaced frequently. The chief causes of deterioration in use are (1) reversible poisoning due to impurities in the reactants or to side-reactions, and (2) irreversible physical changes including loss of surface area (sintering) or mechanical failure. Reversible poisoning may with luck be rectified by simple treatment, such as oxidation or washing, without removing the catalyst from the reactor. To guard against physical changes careful attention has to be given to the strength of the catalyst.

The preferred physical form for a catalyst is determined entirely by the manner in which it is to be used. Its particle size especially is fixed, at least within broad limits, by the type of reactor to be employed. A rough criterion being whether or not one can distinguish separate particles with the naked eye.

When the reactants are solely gases or vapors, there are only two possible basic types of reactor that can be used. A fixed-bed reactor is one in which a tube is packed with "coarse" catalyst particles through which the reactants flow. As a result of the obstruction to gas flow by the particles, a pressure drop occurs across the bed, and a positive pressure has to be applied at the inlet to secure an adequate flow-rate. The size of the pressure drop increases with increasing flow-rate and bed length, and with decreasing particle size. Many different shapes and sizes of catalyst particles are used in fixed-bed reactors, and a great deal of skill and experience goes into choosing the most appropriate for a given set of conditions. Dimensions of more common forms are usually between 2 mm and 2 cm.

When, as in the case of ammonia oxidation, it is desired to use a pure metal catalyst at high temperature and a very short time, a bed of 20 to 30 finely woven metal gauges is used; this is another catalyst form suited for use in a fixed-bed reactor.

Since many catalyzed reactions are strongly exothermic, fixed-bed reactors must contain a facility for removing the unwanted enthalpy of reaction. This may be done in one of two ways. A multi tubular reactor contains a large number of reaction tubes, typically 2.5 cm in diameter, with a cooling gas or liquid flowing between them. Alternatively the bed may be split into sections, with provision for cooling the gas in the spaces between.

A further important and quite different catalyst form is finding application, especially in air-pollution control. This is the so-called monolithic structure, which consists of a block of ceramic material (α-alumina or mullite) through which run fine parallel channels. The structures can be made in many different ways, each of which results in a characteristic channel shape. The blocks are obtainable in a range of shapes and sizes, and a single block when inserted in a container therefore behaves as a fixed-bed reactor. Because of the low porosity of the material from which these structures are made, it is usually necessary to attach a thin layer of a more porous substance to which the catalytically active phase can adhere. Such monolithic structures have several clear advantages over pellets or granules packed loosely in a tube. First, the pressure drop through the bed is less for equal quantities of catalyst, thus permitting higher space velocities to be achieved; secondly, there is no attrition of the catalyst due to particles rubbing against each other and forming fine dust. Their ability to withstand thermal shock is however in some cases limited.

The second basic kind of reactor for gaseous reactions is the fluidized bed. Here the catalyst consists of fine particles. And when the gas flow upwards through a bed of such powder attains a critical velocity the bed appears to "boil": it expands significantly, and the particles are in continuous motion. In this state the bed is said to be fluidized. This behavior gives certain advantages over fixed beds: for examples, heat transfer characteristics are much better, and the pressure drop increases far less quickly with increasing flow-rate.

Words and Expressions

desired reaction	目标反应
irreversible physical change	不可逆物理变化
fixed-bed reactor	固定床反应器
tubular reactor	管式反应器
catalyst particles	催化剂颗粒
space velocity	空速
critical velocity	临界速度

Catalyst

Notes

1. Because of the low porosity of the material from which these structures are made, it is usually necessary to attach a thin layer of a more porous substance to which the catalytically active phase can adhere.

译文：由于材料的低孔隙率结构，所以它通常需要附着一薄层多孔物质来黏附催化活性相。

2. And when the gas flow upwards through a bed of such powder attains a critical velocity the bed appears to "boil": it expands significantly, and the particles are in continuous motion.

译文：当气体向上流动通过床层，粉末达到临界速度表现为沸腾：粉末明显膨胀，颗粒连续运动。

Reading Comprehension

1. What are the most considerations of the industrially successful catalyst?
2. How many basic kinds of reactors for gaseous reactions are there?

Reading Material

Phase-transfer Catalyst 相转移催化剂

In chemistry, a phase-transfer catalyst or PTC is a catalyst that facilitates the migration of a reactant from one phase into another phase where reaction occurs. Phase-transfer catalysis is a special form of heterogeneous catalysis. Ionic reactants are often soluble in an aqueous phase but insoluble in an organic phase in the absence of the phase-transfer catalyst. The catalyst functions like a detergent for solubilizing the salts into the organic phase. Phase-transfer catalysis refers to the acceleration of the reaction upon the addition of the phase-transfer catalyst.

By using a PTC process, one can achieve faster reactions, obtain higher conversions or yields, make fewer byproducts, eliminate the need for expensive or dangerous solvents that will dissolve all the reactants in one phase, eliminate the need for expensive raw materials and/or minimize waste problems. Phase-transfer catalysts are especially useful in green chemistry — by allowing the use of water, the need for organic solvents is reduced.

Contrary to common perception, PTC is not limited to systems with hydrophilic and hydrophobic reactants. PTC is sometimes employed in liquid/solid and liquid/gas reactions. As the name implies, one or more of the reactants are transported into a second phase which contains both reactants. Phase-transfer catalysts for anionic reactants are often quaternary ammonium and phosphonium salts. Typical catalysts include benzyltrimethylammonium chloride and hexadecyltributylphosphonium bromide.

For example, the nucleophilic aliphatic substitution reaction of an aqueous sodium cyanide solution with an ethereal solution of 1-bromooctane does not readily occur. The 1-bromooctane is poorly soluble in the aqueous cyanide solution, and the sodium cyanide does not dissolve well in the ether. Upon the addition of small amounts of hexadecyltributylphosphonium bromide, a rapid reaction ensues to give nonyl nitrile:

$C_8H_{17}Br(org) + NaCN(aq) \longrightarrow C_8H_{17}CN(org) + NaBr(aq)$ (catalyzed by a R4P+Cl− PTC)

Via the quaternary phosphonium cation, cyanide ions are "ferried" from the aqueous phase into the organic phase.

Subsequent work demonstrated that many such reactions can be performed rapidly at around room temperature using catalysts such as tetra-n-butylammonium bromide or methyltrioctylammonium chloride in benzene/water systems. An alternative to the use of "quat salts" is to convert alkali metal cations into hydrophobic cations. In the research lab, crown ethers are used for this purpose. Polyethylene glycols are more commonly used in practical applications. These ligands encapsulate alkali metal cations (typically Na^+ and K^+), affording large lipophilic cations. These polyethers have a hydrophilic "interiors" containing the ion and a hydrophobic exterior.

PTC is widely exploited industrially. Polyester polymers for example are prepared from acid chlorides and bisphenol-A. Phosphothioate-based pesticides are generated by PTC- catalyzed alkylation of phosphothioates. One of the more complex applications of PTC involves asymmetric alkylations, which are catalyzed by chiral quaternary ammonium salts derived from cinchona alkaloids.

History, Inheritance and Development

Hou's Process

This process was developed by Chinese chemist Hou Debang in the 1930s. The earlier steam reforming byproduct carbon dioxide was pumped through a saturated solution of sodium chloride and ammonia to produce sodium bicarbonate by these reactions:

$$CH_4 + 2H_2O \longrightarrow CO_2 + 4H_2$$
$$3H_2 + N_2 \longrightarrow 2NH_3$$
$$NH_3 + CO_2 + H_2O \longrightarrow NH_4HCO_3$$
$$NH_4HCO_3 + NaCl \longrightarrow NH_4Cl + NaHCO_3$$

The sodium bicarbonate was collected as a precipitate due to its low solubility and then heated up to approximately 80 °C (176 °F) or 95 °C (203 °F) to yield pure sodium carbonate similar to last step of the Solvay process. More sodium chloride is added to the remaining solution of ammonium and sodium chlorides; also, more ammonia is pumped at 30-40 °C to this solution. The solution temperature is then lowered to below 10 °C. Solubility of ammonium chloride is higher than that of sodium chloride at 30 °C and lower at 10 °C. Due to this temperature-dependent solubility difference and the common-ion effect, ammonium chloride is precipitated in a sodium chloride solution.

The Chinese name of Hou's process, lianhe zhijian fa (联合制碱法), means "coupled manufacturing alkali method": Hou's process is coupled to the Haber process and offers better atom economy by eliminating the production of calcium chloride, since ammonia no longer needs to be regenerated. The byproduct ammonium chloride can be sold as a fertilizer.

Practice and Training

WorldSkills

WorldSkills Kazan 2019

WorldSkills organises the world championships of vocational skills, and is held every two years in different parts of the world. The organisation, which also hosts conferences about vocational skills, describes itself as the global hub for skills.

WorldSkills currently has 85 Member countries and regions, most of which organise national skills competitions that help to prepare the workforce and talent of today for the jobs of the future.

WorldSkills International, formerly known as the International Vocation Training Organisation (IVTO), was founded in the 1940s and emerged from a desire to create new employment opportunities for young people in some of the economies that were devastated by the Second World War.

Future Competitions

WorldSkills Competition 2022 Special Edition

WorldSkills competition 2022 will be held in 15 countries and regions between 7 September and 26 November 2022. The event was originally scheduled for September 2021, but due to the COVID-19 global pandemic, it was decided to shift the event to 2022. Thanks to the commitment of our Partners and Members, 62 skill competitions will be held over 12 weeks.

WorldSkills Lyon 2024

Lyon, France was selected as the host city for the 47th WorldSkills competition. Originally scheduled for 12-17 September 2023, it has been moved back one year due to the decision to postpone by one year the 46th WorldSkills event originally scheduled to be held in 2021 due to the COVID-19 global pandemic. This will be the second time that France has hosted the WorldSkills competition, the first time being in 1995.

2019 Medal Table

Rank	Nation	Gold	Silver	Bronze	Total
1	China(CHN)	16	14	5	35
2	Russia(RUS)	14	4	4	22
3	South Korea(KOR)	7	6	2	15

WorldSkills
Occupational Standards

Unit Six

Fine Chemicals

精细化学品

Lesson One A Brief Introduction of Fine Chemicals
精细化学品简介

A brief introduction of fine chemicals

Until the early 1970s in-house capabilities for production of raw materials and intermediates for products sold were considered a key competitive advantage by the chemical industry. As of this writing, this situation has changed completely. Particularly those chemical companies concentrating on portfolios having high added value specialties consider efficient research and development, dynamic marketing, and proper management of human, technical, and financial resources as key success factors rather than production. This change in strategic focus is especially evident in the agrochemical and pharmaceutical industries which together comprise the life science industry. In manufacturing activities, manufacturing has been regrouped into separate divisions, and in a few cases large life science companies have disinvested their chemical manufacturing activities. In addition to these strategic developments, the requirement for more and more sophisticated organic chemical has contributed substantially to the emergence of the fine chemicals industry as a distinct entity. Fine chemical manufacturers are backward integrated, production oriented, and service the mega enterprises within the chemical industry. The fine chemicals industry has its own characteristics with regard to research and development, production, marketing, and finance.

In the chemical business products may be described as commodities, fine chemicals, of specialists. Various commodities are also known as petrochemicals, basic chemicals, organic chemicals (large-volume), monomers, commodity fibers, and plastics. Advanced intermediates, building blocks, bulk drugs and bulk pesticides, active ingredients, bulk vitamins, and flavor and fragrance chemicals are all fine chemicals. Adhesives, diagnostics, disinfectants, electronic chemicals, food additives, mining chemicals, pesticides, pharmaceuticals, photographic chemicals, specialty polymers, and water treatment chemicals are all specialties. The added value is highest for specialties.

It is common to both commodities and fine chemicals that these materials are identified according to specifications, according to what they are. These substances are sold within the chemistry industry, and customers know better how to use them than suppliers. Specialties are identified according to performance, according to what they can do. Customers are the public, and suppliers have to provide for technical assistance. A particular substance may be both a fine chemical and a specialty. For example, as long as, 2-chloro-5(1-hydroxy-3-oxo-1-isoindolinyl) benzene sulfonamide is sold according to specifications it is a fine chemical. But once it is tabletted and marked as the diuretic chlorthalidone, it becomes a specialty. The limits between commodities and fine chemicals are not so clearly fixed.

In terms of volume the border line between commodities and fine chemicals comes somewhere between about 1,000t/yr and 10,000t/yr. In terms of unit prices the line typically varies between $2.50/kg and $10/kg. Establishing more precise demarcations is not practical even though a large number of well-known intermediates fall within these lines, e.g. acetoacetanilide, chloroformate, cyanuric chloride, hydroquinone, malonate, pyridine, picoline, and sorbic acid. Additionally, for amino acids and vitamins two typical groups of fine chemicals, the two largest volume products, L-lysine, ascorbic acid and niacin, respectively, are sold in quantities exceeding 10,000t/yr. The prices range beyond the $10/kg level as well.

Products

If fine chemicals are classified according to applications, the most prominent categories in terms of tonnage volumes are the ones used in the production for agrochemicals, followed by pharmaceutical fine chemicals. Within agrochemicals, triazine herbicides, from the key intermediate cyanuric chloride, are produced in quantities exceeding 100,000 metric tons per annum. Chloroacetanilide, from the key intermediates 2,6-diethylaniline and chloroacetylchloride, and phenoxy herbicides, from L-2-chloropropanoic acid rank between 50,000t/yr and 100,000 t/yr. Also phosgene-derived thiocarbamate and urea herbicides, as well as dithiocarbamate fungicides are very large-volume products. However, most of the constituents of fungicides are commodities. This is also true for organophosphate insecticides derived from phosphorous oxychloride. All of these agrochemicals show zero or even negative growth, because of gradual replacement by more active, lower volume crop protection chemicals. Well-known examples are sulfonyl ureas, many of which contain 2-amino-4, 6-dissubstituted pyrimidine moieties and imidazolines, the key intermediate of antibiotics which pyridine-2, 3-dicarboxylic acid.

Within pharmaceuticals, the highest volume categories are vitamins painkillers and β-lactam antibiotics has a world production of some 50,000 t/yr; ascorbic acid (vitamin C) totals 50,000t/yr; and niacin (vitamin PP) totals some 25,000 t/yr. There are large production volumes of β-lactam antibiotic precursors such as 6-aminopenicillinic acid and 7-aminocephalosporanic acid, as well as side chains D-phenylglycine and D-p-hydroxyphenylglycine, developed for the first semisynthetic penicillins, manufactured in the multithousand metric tons per year range. 2-Aminothiazolyl alkoximinoacetates, used for thirdgeneration cephalosporins, are made in the several several hundred metric ton per annum, as are five-ring heterocycles derived from sodium azide such as 5-mercapto-1-methyltetrazole and 2-mercapto-5-methyl-1,3,4-thiadiazole.

In terms of types molecular structures, heterocyclic compounds are the most important fine chemical category, especially fine chemicals having an N-heterocyclic structure as found in the vitamins B2, B6, H, PP and folic acid. These and other natural substances have gained a great importance in modern pharmaceuticals and pesticides. Even modern pigments and engineering plastics exhibit an N-heterocyclic structure.

From the point of view of application, pharmaceutical fine chemicals constitute the largest part of all fine chemicals, both in terms of number of products and volume of sales. About 40%-50% of the total fine chemicals sales come from pharmaceutical fine chemicals; about 20%-25% are agrochemicals, and the rest belong to other categories.

Not many fine chemicals have a production value exceeding $10 million per year. Less than a dozen achieve production volumes about 10,000 metric tons per year and sales of above $100 million per year. Apart from the pharmaceutical and pesticide fine chemicals these comprise the amino acids L-lysine and D, L-methionine used as feed additives, and vitamins ascorbic acid and nicotinic acid.

The future development of the fine chemicals industry depends mainly on the development of demand. The growth of the fine chemicals business is mainly fostered by the introduction of new pharmaceuticals, agrochemicals, engineering plastics, and other specialties requiring high value organic intermediates.

Words and Expressions

portfolio [pɔːtˈfəuljəu]	n. 业务量，业务责任
agrochemical [ˌægrəuˈkemɪkl]	n. 农用化学品，用农产品制得的化学品
pharmaceutical [ˌfɑːməˈsjuːtɪkəl]	adj. 制药（学）上的
disinvest [dɪsɪnˈvest]	v.（常与 from 连用）减资，投资缩减
entity [ˈentɪtɪ]	n. 实体
mega [ˈmegə]	n. 百万，大
fine chemicals	精细化学品
commodity [kəˈmɒdɪtɪ]	n. 日用品
pesticide [ˈpestɪsaɪd]	n. 杀虫剂
according to	依照
as long as	只要，在……的时候
hydroxyl [haɪˈdrɒksɪl]	n. [化] 羟（基），氢氧基，氢氧化物
isoindolinyl [aɪsɔɪndəˈlɪnɪl]	n. 异二氢氮杂茚基
sulfonamide [sʌlˈfɒnəmaɪd]	n. 磺胺
tablet [ˈtæblɪt]	v. 把……压成片（块），制片（块）
diuretic [ˌdaɪjuəˈretɪk]	adj. 利尿的
chlorthalidone [klɔːˈθælɪdoun]	n. [药]氯噻酮（利尿降压药）
acetoacetanilide [æsɪtəæsɪˈtænɪlaɪd]	n. [化]N-乙酰乙酰苯胺，N-丁间酮酰苯胺
chloroformate [klɔːrəˈfɔːmeɪt]	n. [化]氯甲酸酯
malonate [mæloʊˈneɪt]	n. [化]丙二酸
picoline [ˈpaɪkəˌliːn]	n. [化]皮考啉，甲基吡啶
triazine [ˈtraɪəˌziːn]	n. [化]三嗪

imidazoline [ɪmɪdə'liːn]　　　　　　　　n. 咪唑啉
azide ['æzaɪd]　　　　　　　　　　　　n. [化]叠氮化物
methionine [məˈθaɪəˌniːn]　　　　　　 n. [生化]蛋氨酸，甲硫氨酸

Notes

1. Particularly those chemical companies concentrating on portfolios having high added value specialties consider efficient research and development, dynamic marketing, and proper management of human, technical, and financial resources as key success factors rather than production.

句子分析：分词短语 concentrating on 修饰 companies；分词短语 having high added value specialties 修饰 portfolios（公事包，工作，事务）。

译文：尤其是一些集中在高附加值专用化学品方面的公司认为，高效的研究与开发，充满活力的市场以及对人力、技术及财力资源的合理利用，是比生产更为重要的成功因素。

2. It is common to both commodities and fine chemicals that these materials are identified according to specifications, according to what they are.

句子分析：what they are 是 according to 的宾语，that these materials are identified 为 It is common to both commodities and fine chemicals 引导的主语从句。

译文：这一点对大宗类化学品与精细化学品来说是共同的，即这两类都是根据它们的产品规格指标，即它们的内在属性进行分类的。

Reading Comprehension

1. Is the definition of fine chemicals?
2. What are the fine chemicals around us?

Reading Material

Introduction of Organic Synthesis　有机合成简介

Organic synthesis is a special branch of chemical synthesis and is concerned with the construction of organic compounds via organic reactions. Organic molecules often contain a higher level of complexity than purely inorganic compounds, so that the synthesis of organic compounds has developed into one of the most important branches of organic chemistry. There are several main areas of research within the general area of organic synthesis: total synthesis, semisynthesis, and methodology.

Total synthesis is the complete chemical synthesis of an organic molecule from simpler pieces called precursors. It usually refers to a process not involving the aid of biological processes, which distinguishes it from semisynthesis. Since there are usually many steps in total synthesis, the chemist must develop a plan, called a route, or adopt one already known. There may be several different or partially different routes to the target including the choice of the substrates. To be commercially viable, syntheses must use readily obtainable, bulk quantities. Commercial synthesis often relies on petrochemical precursors. Sometimes, however, the chemist proceeds on a small scale. The target molecules can be natural products, medicinally important active ingredients, or organic compounds of theoretical interest. Often the aim is to discover new route of synthesis for a target molecule for which there already exist known routes. Sometimes no route exists and the chemist wishes find a viable route for the first time.

Semisynthesis or partial chemical synthesis is a type of chemical synthesis that uses compounds isolated from natural sources (e.g. plant material or bacterial or cell cultures) as starting materials. These natural biomolecules are usually large and complex molecules. This is opposed to a total synthesis where large molecules are synthesized from a stepwise combination of small and cheap (usually petrochemical) building blocks. Semisynthesis is usually used when the precursor molecule is too structurally complex, too costly or too inefficient to be produced by total synthesis. From a synthesis viewpoint, life is an amazing chemical factory, and for some chemical products, it is hard to compete with life as a synthetic route—sometimes a lowly plant can grow for pennies' worth of amortized farming investment what total synthesis would struggle expensively to produce. Thus humanity collaborates with nature to make the most of this remarkable cost-effectiveness.

Drugs derived from natural sources are usually produced by harvesting the natural source or through semisynthetic methods: one example is the semisynthesis of LSD from ergotamine, which is isolated from ergot fungus cultures. The commercial production of paclitaxel is also based on semisynthesis.

The antimalarial drug artemether is a semisynthetic derived from naturally occurring artemisinin. The latter is unstable due to the presence of a lactone group and therefore this group is replaced by an acetal through organic reduction with potassium borohydride and methoxylation:

Methodology and applications: Each step of a synthesis involves a chemical reaction, and reagents and conditions for each of these reactions must be designed to give an adequate yield of pure product, with as little work as possible. A method may already exist in the literature for making one of the early synthetic intermediates, and this method will usually be used rather than an effort to "reinvent the wheel". However, most intermediates are compounds that have never been made before, and these will normally be made using general methods developed by methodology researchers. To be useful, these methods need to give high yields, and to be reliable for a broad range of substrates. For practical applications, additional hurdles include industrial standards of safety and purity. Methodology research usually involves three main stages: discovery, optimisation, and studies of scope and limitations. The discovery requires extensive knowledge of and experience with chemical reactivities of appropriate reagents. Optimisation is a process in which one or two starting compounds are tested in the reaction under a wide variety of conditions of temperature, solvent, reaction time, etc., until the optimum conditions for product yield and purity are found. Finally, the researcher tries to extend the method to a broad range of different starting materials, to find the scope and limitations.

Lesson Two   Coating
涂料

Coating

Coating is a covering that is applied to the surface of an object, usually referred to as the substrate. In many cases coatings are applied to improve surface properties of the substrate, such as appearance, adhesion, wetability, corrosion resistance, wear resistance, and scratch resistance. In other cases, in particular in printing processes and semiconductor device fabrication (where the substrate is a wafer), the coating forms an essential part of the finished product.

Coatings may be applied as liquids, gases or solids. Coatings can be measured and tested for proper opacity and film thickness by using a Drawdown card.

Enamel (paint)

An enamel paint is a paint that air dries to a hard, usually glossy, finish. In reality, most commercially-available enamel paints are significantly softer than either vitreous enamel or stoved synthetic resins.

Some enamel paints have been made by adding varnish to oil-based paint.

Typically the term "enamel paint" is used to describe oil-based covering products, usually with a significant amount of gloss in them, however recently many latex or water-based paints have adopted the term as well. The term today means "hard surfaced paint" and usually is in reference to paint brands of higher quality, floor coatings of a high gloss finish, or spray paints.

Powder coating

Powder coating is a type of coating that is applied as a free-flowing, dry powder. The main difference between a conventional liquid paint and a powder coating is that the powder coating does not require a solvent to keep the binder and filler parts in a liquid suspension form. The coating is typically applied electrostatically and is then cured under heat to allow it to flow and form a "skin". The powder may be a thermoplastic or a thermoset polymer. It is usually used to create a hard finish that is tougher than conventional paint. Powder coating is mainly used for coating of metals, such as "whiteware", aluminium extrusions, and automobile and bicycle parts. Newer technologies allow other materials, such as MDF (medium-density fibreboard), to be powder coated using different methods.

Industrial coating

An industrial coating is a paint or coating defined by its protective, rather than its aesthetic properties, although it can provide both.

The most common use of industrial coatings is for corrosion control of steel or concrete. Other functions include intumescent coatings for fire resistance. The most common polymers used in industrial coatings are polyurethane, epoxy and moisture cure urethane. Another highly common polymer used in industrial coating is a fluoropolymer. There are many types of industrial coatings including inorganic zinc, phosphate, and Xylan and PVD coatings.

NACE International and The Society for Protective Coatings (SSPC) are professional organizations involved in the industrial coatings industry. Industrial coatings are often composites of various substances. For example the Xylan (a trade mark of Whitford Worldwide)line of dry-film lubricants are composites of fluoropolymers (typically PTFE, PFA, and FEP) and reinforcing thermoset polyimide and polyamide binder resins initially suspended in a variety of solvents (such as ethyl acetate, xylene, dimethylformamide, N-methyl 2-pyrrolidone, etc.).

A complete coating system may include a primer, the coating, and a sealant/top-coat.

Silicate mineral paint

Mineral paints are mineral based coatings formulated with potassium silicate or sodium silicate, otherwise known as waterglass as the binder, combined with inorganic, alkaline-resistant pigments. They are fully inorganic (containing no organic solvents) and are non-offgassing. Mineral paints petrify, by binding to any silicates within the substrate, forming a micro-crystalline structure and a breathable finish. They are more of a stain, which becomes integral to the substrate, rather than a coating. They are alkaline and therefore inhibit microbiotic growth, and reduce carbonization of cementitious materials.

The majority of non-toxic concrete stains and limestone restoration products are waterglass based. Mineral paints are also used as a non-toxic wood preservative.

The difference between the use of sodium silicate and potassium silicate as a binder is mainly geographic. The western hemisphere mainly produces sodium silicate, where Europe produces potassium silicate.

Fusion bonded epoxy coating (FBE coating)

Fusion bonded epoxy coating, also known as fusion-bond epoxy powder coating and commonly referred to as FBE coating, is an epoxy based powder coating that is widely used to protect steel pipe used in pipeline construction, concrete reinforcing bars (rebar) and on a wide variety of piping connections, valves etc. from corrosion. FBE coatings are thermoset polymer coatings. They come under the category of 'protective coatings' in paints and coating nomenclature. The name 'fusion-bond epoxy' is due to resin cross-linking and the application method, which is different from a conventional paint. The resin and hardener components in the dry powder FBE stock remain unreacted at normal storage conditions. At typical coating application temperatures, usually in the range of 180℃ to 250℃ (356°F to 482°F), the contents of the powder melt and transform to a liquid form. The liquid FBE film wets and flows onto the steel surface on which it is applied, and soon becomes a solid coating by chemical cross-linking, assisted by heat. This process is known as "fusion bonding". The chemical cross-linking reaction taking place in this case is irreversible. Once the curing takes place, the coating cannot be returned to its original form by any means. Application of further heating will not "melt" the coating and thus it is known as a "thermoset" coating. The world's leading FBE manufacturers are Valspar, KCC Corporation, Jotun Powder Coatings, 3M, DuPont, Akzo Nobel, BASF and Rohm & Haas.

Words and Expressions

substrate ['sʌbstreɪt]	n. （供绘画、印刷等的）底面，基底，基片
scratch [skrætʃ]	n. 抓，搔痒，抓痕，抓伤，刮擦声
wafer ['weɪfə]	n. 晶片，圆片
enamel [ɪ'næməl]	n. 搪瓷，珐琅，釉药，瓷漆
glossy ['glɒsɪ]	adj. 光洁的，光滑的
vitreous ['vɪtrɪəs]	adj. 玻璃（似）的，玻璃质的
aesthetic [iːs'θetɪk]	adj. 有关美的，审美的，悦目的，雅致的
xylan ['zaɪlæn]	n. 木聚糖
petrify ['pɛtrəˌfaɪ]	n. 吓呆，使麻木
limestone ['laɪmstoʊn]	n. 石灰石
fusion ['fjuːʒən]	n. 熔合，熔接，融合，核聚变，联合，合并

Notes

1. Coatings may be applied as liquids, gases or solids. Coatings can be measured and tested for proper opacity and film thickness by using a Drawdown card.

译文：涂料可以是液体、气体或固体。涂料可以通过刮涂卡片测试其不透明性和膜厚。

2. The most common polymers used in industrial coatings are polyurethane, epoxy and moisture cure urethane.

译文：在工业涂料中最常见的聚合物有聚氨酯、环氧树脂、湿气固化聚氨酯。

3. The difference between the use of sodium silicate and potassium silicate as a binder is mainly geographic. The western hemisphere mainly produces sodium silicate, where Europe produces potassium silicate

译文：硅酸钠和硅酸钾主要是根据地理位置的不同而选择使用。西半球（美加等国）主要用硅酸钠，欧洲主要用硅酸钾。

Reading Comprehension

1. Point out the differences between the coatings and the paints.
2. Write out the curing mechanism of coatings.
3. Do you know any famous paint companies?

Reading Material

Dyes 染料

The most common dyes are the azo dyes, formed by coupling diazotized amines to phenols. The dye can be made in bulk, or, as we shall see, the dye molecule can be developed on and in the fiber by combining the reactants in the presence of the fiber.

One dye, Orange II, is made by coupling diazotized sulfanilic acid with 2-naphthol in alkaline solution; another, Methyl Orange, is prepared by coupling the same diazonium salt with N,N-dimethylaniline in a weakly acidic solution. Methyl Orange is used as an indicator as it changes color at pH 3.2-4.4. The change in color is due to transition from one chromophore (azo group) to another (quinonoid system).

You are to prepare one of these two dyes and then exchange samples with a neighbor and do the tests with both dyes. Both substances dye wool, silk, and skin, and you must work carefully to avoid getting them on your hands or clothes. The dye will eventually wear off your hands or they can be cleaned by soaking them in warm, slightly acidic (H_2SO_4) permanganate solution until heavily stained with manganese dioxide and then removing the stain in a bath of warm, dilute bisulfite solution.

Experiments
Diazotization of Sulfanilic Acid

In a 125-mL Erlenmeyer flask dissolve, by boiling, 4.8 g of sulfanilic acid monohydrate in 50 mL of 2.5% sodium carbonate solution (or use 1.33 g of anhydrous sodium carbonate and 50 mL of water). Cool the solution under the tap, add 1.9 g of sodium nitrite, and stir until it is dissolved. Pour the solution into a flask containing about 25 g of ice and 5 mL of concentrated hydrochloric acid. In a minute or two a powdery white precipitate of the diazonium salt should separate and the material is then ready for use. The product is not collected but is used in the preparation of the dye Orange II and/or Methyl Orange while in suspension. It is more stable than most diazonium salts and will keep for a few hours.

Orange II (1-*p*-Sulfobenzeneazo-2-Naphthol Sodium Salt

In a 400-mL beaker dissolve 3.6 g of 2-naphthol in 20 mL of cold 10% sodium hydroxide solution and pour into this solution, with stirring, the suspension of diazotized sulfanilic acid from Section 2. Rinse the Erlenmeyer flask with a small amount of water and add it to the beaker. Coupling occurs very rapidly and the dye, being a sodium salt, separates easily from the solution because a considerable excess of sodium ion from the carbonate, the nitrite, and the alkali is present. Stir the crystalline paste thoroughly to effect good mixing and, after 5-10 min, heat the mixture until the solid dissolves. Add 10 g of sodium chloride to further decrease the solubility of the product, bring this all into solution by heating and stirring, set the beaker in a pan of ice and water, and let the solution cool undisturbed. When near room temperature, cool further by stirring and collect the product on a Buchner funnel. Use saturated sodium chloride solution rather than water for rinsing the material out of the beaker and for washing the filter cake free of the dark-colored mother liquor. The filtration is somewhat slow.

The product dries slowly and it contains about 20% of sodium chloride. The crude yield is thus not significant, and the material need not be dried before being purified. This particular azo dye is too soluble to be crystallized from water; it can be obtained in a fairly satisfactory form by adding saturated sodium chloride solution to a hot, filtered solution in water and cooling, but the best crystals are obtained from aqueous ethanol. Transfer the filter cake to a beaker, wash the material from the filter paper and funnel with water, and bring the cake into solution at the boiling point. Avoid a large excess of water, but use enough to prevent separation of solid during filtration (use about 50 mL). Filter by suction through a Buchner funnel that has been preheated on the steam bath. Pour the filtrate into an Erlenmeyer flask, rinse the filter flask with a small quantity of water, add it to the flask, estimate the volume, and if greater than 60 mL evaporate by boiling. Cool to 80℃, add 100-125 mL of ethanol, and allow crystallization to proceed. Cool the solution well before collecting the product. Rinse the beaker with mother liquor and wash finally with a little ethanol. The yield of pure, crystalline material is 6.8 g. Orange II separates from aqueous alcohol with two molecules of water of crystallization and allowance for this should be made in calculating the yield. If the water of hydration is eliminated by drying at 120℃ the material becomes fiery red.

Cleaning Up

The filtrate from the reaction, although highly colored, contains little dye, but is very soluble in water. It can be diluted with a large quantity of water and flushed down the drain or, with the volume kept as small as possible, it can be placed in the aromatic amines hazardous waste container or it can be reduced with tin(II) chloride. The crystallization filtrate should go into the organic solvents container.

Lesson Three Classification and Application of Surfactants
表面活性剂的分类与应用

Types of surfactants

Numerous variations are possible within the structure of both the head and tail group of surfactants. The head group can be charged or neutral, small and compact in size, or polymeric chain. The tail group is usually a single or double, straight or branched hydrocarbon chain, but may also be fluorocarbon, or a siloxane, or contain aromatic group. Commonly encountered hydrophilic and hydrophobic groups are listed in Table 6-1 and Table 6-2 respectively.

Classification and application of surfactants

Since the hydrophilic part normally achieves its solubility either by ionic interactions or by hydrogen bonding, the simplest classification is based on surfactant head group type, with further subgroups according to the nature of the lyophobic moiety. Four basic classes therefore emerge as:

The anionics and cationics, which dissociate in water into two oppositely charged species (the surfactant ion and its counterion),

The non-ionics, which includes a highly polar (non charged) moiety, such as polyoxyethylene (—OCH$_2$CH$_2$O—) or polyol groups,

And the amphoterics, it combines both a positive group and a negative group.

With the continuous search for improving surfactant properties, new structure have recently emerged that exhibit interesting synergistic interactions or enhanced surface and aggregation properties. These novel surfactants have attracted much interest and include catanionics, bolaforms, gemini (or dimeric) surfactants, polymeric and polymerisable surfactants. Characteristics and typical examples are shown in Table 6-1. Another important driving force for this research is the need for enhanced surfactant biodegradability In particular for personal care products and household detergents, regulations require high biodegradability and non-toxicity of each component in the formulation.

Table 6-1 Common hydrophilic groups found in commercially available surfactants

Class	General structure
sulfonate	R—SO$_3^-$M$^+$
sulfate	R—OSO$_3^-$M$^+$
carboxylate	R—COO$^-$M$^+$
phosphate	R—OPO$_3^-$M$^+$
ammonium	R$_x$H$_y$N$^+$ X$^-$ (x = 1~3, y = 4−x)
quaternary	R$_4$N$^+$ X$^-$
betaines	R$_4$N$^+$(CH$_3$)$_2$CH$_2$COO$^-$
sulfobetianes	RN$^+$(CH$_3$)$_2$CH$_2$CH$_2$SO$_3^-$
polyoxyethylene(POE)	R—OCH$_2$CH$_2$(OCH$_2$CH$_2$)$_n$O
polyols	sucrose, sorbitan, glycerol, ethylene glycol, etc.
polypeptide	R—NH—CHR—CO—NH—CHR'—CO—⋯—COOH
polyglycidyl	R—(OCH$_2$CH[CH$_2$OH]CH$_2$)$_n$—⋯—OCH$_2$CH[CH$_2$OH]CH$_2$OH

Table 6-2 Common hydrophobic groups used in commercially available surfactants

Group	General structure	
natural fatty acids	$CH_3(CH_2)_nCOOH$	$n=12\sim18$
petroleum paraffins	$CH_3(CH_2)_nCH_3$	$n=8\sim10$
olefins	$CH_3(CH_2)_nCH=CH_2$	$n=7\sim17$
alkylbenzene	$CH_3(CH_2)_nCH_2-\phi$	$n=6\sim10$, linear or branched
alkylaromatics	$CH_2(CH_2)_nCH_3$ substituted naphthalene with R groups	$n=1\sim2$ for water soluble, $n=8$ or 9 for oil soluble surfactants
alkylphenols	$CH_3(CH_2)_nCH_2-\phi-OH$	$n=6\sim10$, linear or branched

Surfactant uses and development

Surfactants may be from natural or synthetic sources. The first category includes naturally occurring amphiphiles such as the lipids, which are surfactants based on glycerol and are vital components of the cell membrane. Also in this group are the so-called "soaps", the first recognized surfactants. These can be traced back to Egyptian times; by combining animal and vegetable oils with alkaline salts a soap-like material was formed, and this was used for treating skin diseases, as well as for washing. Soaps remained the only source of natural detergents from the seventh century till the early twentieth century, with gradually more varieties becoming available for shaving and shampooing, as well as bathing and laundering. In 1916, in response to a World War Ⅰ-related shortage of fats or making soap, the first synthetic detergent was developed in Germany. Known today simply as detergents, synthetic detergents are washing and cleaning products obtained from a variety of raw materials.

Nowadays, synthetic surfactants are essential components in many industrial processes and formulations. Depending on the precise chemical nature of the product, the properties of, for example emulsification, detergency and forming may be exhibited in varying degree. The number and arrangement of the hydrocarbon groups together with the nature and position of the hydrophilic groups combine to determine the surface-active properties of the molecule. For example C_{12} to C_{20} is generally regarded as the range covering optimum detergency, whilst wetting and foaming are best achieved with shorter chain lengths. Structure performance relationships and chemical compatibility are therefore key element in surfactant-based formulations, so that much research is devoted to this area.

Among the different classes of surfactants, anionics are often used in greater volume than any other types, mainly because of the ease and low cost of manufacture. They contain negatively charged head group, e.g. carboxylates ($-COO-$), used in soaps, sulfate ($-OSO_3-$), and sulfonates ($-SO_3-$) groups. Their main applications are in detergency, personal care products, emulsifers and soaps.

Cationics have positively charged head groups —e.g. trimethylammonium ion and are mainly involved in applications related to their absorption at surfaces. These are generally negative charged (e.g. metal, plastics, minerals, fibers, hairs and cell membranes) so that they can be modified upon treatment with cationic surfactants. They are therefore used as anticorrosion and antistatic agents, flotation collectors, fabric softeners, hair conditioners and bactericides.

Non-ionics contain groups with a strong affinity for water due to strong dipole-dipole interactions arising from hydrogen bonding, e.g. ethoxylates. One advantage over ionics is that the length of both the hydrophilic and hydrophobic groups can be varied to obtain maximum efficiency in use. They find applications in low temperature detergents and emulsifiers.

Zwitterionics constitute the smallest surfactant class due to their high cost of manufacture. They are characterized by excellent dermatological properties and skin compatibility. Because of their low eye and skin irritation, common uses are in shampoos and cosmetics.

Words and Expressions

surfactant [sɜːrˈfæktənt]	n. 表面活性剂；adj. 表面活性剂（的）
siloxane [saɪˈlɒkˌseɪn]	n. 硅氧烷
lyophobic [ˌlaɪəˈfoʊbɪk]	adj. 疏水的
anionic [ˈænaɪənɪk]	adj. 阴离子的
cationic [ˈkætaɪənɪk]	adj. 阳离子的
polyoxyethylene [ˈpoʊljɒksjeθɪliːn]	n. 聚乙氧基
amphoteric [ˌæmfəˈterɪk]	adj. 两性的
carboxylate [kɑːˈbɒksɪleɪt]	n. 羧酸盐，羧酸酯
sulfate [ˈsʌlfeɪt]	n. 硫酸盐
sulfonate [ˈsʌlfəˌneɪt]	n. 磺酸盐
emulsifier [ɪˈmʌlsɪfaɪə]	n. 乳化剂
trimethylammonium	n. 三甲基铵化物
zwitterionics [ˈtsvɪtəraɪənɪks]	adj. 两性离子的

Notes

1. Since the hydrophilic part normally achieves its solubility either by ionic interactions or by hydrogen bonding, the simplest classification is based on surfactant head group type, with further subgroups according to the nature of the lyophobic moiety.

译文：由于表面活性剂的亲水部分是通过离子键或氢键的作用而获得水溶性的，因此最简单的分类是根据表面活性剂的亲水部分分类，然后根据其憎水基的性质再进一步分类。

2. Because of their low eye and skin irritation, common uses are in shampoos and cosmetics.

译文：两性离子表面活性剂由于其对眼睛和皮肤的危害较小，因此广泛用于洗发精和化妆品中。

Reading Comprehension

1. What is the definition of surfactants?
2. Write out the typical examples of four surfactants.

Reading Material

Raw Materials for the Manufacture of Soaps

A good quality soap is made from than one and contains more than one non-fatty substances. Good quality soap has to maintain fairly good detergent properties under varying temperatures, with different substance, and in waters of varying hardness.

Following raw materials are required for the manufacture of soaps:

1. Alkalis

Important alkalis used are sodium carbonate, caustic soda, caustic potash, ammonium hydroxide and ethanolamines.

(1) Soda ash or sodium carbonate($Na_2CO_3 \cdot 10H_2O$)

This is non saponifying alkali and is, therefore, used in the preparation of soaps from fatty acids and not from fatty oils.

(2) Caustic soda or sodium hydroxide

This is most widely used alkali in the manufacture of hard soaps which are the most common. Caustic soda, is available in the form of flakes or lumps packed in cylindrical iron drums. A good commercial grade of caustic soda should contain minimum of 96% NaOH, the balance being made up by water, sodium carbonate and sodium chloride.

(3) Caustic potash or potassium hydroxide

Caustic potash is used only for the manufacture of soft soaps. Consumption of this alkali in soap manufacture is very limited.

(4) Ammonium hydroxide

Ammonium hydroxide is available as a concentrated solution(liquid ammonia). It is used with fatty acids, as it cannot saponify oils.

(5) Ethanolamines

These are amino-alcohols and may be considered derivatives of ammonia in which at least one of the hydrogen atoms is replaced by 2-hydroxy ethyl radicals. Several mono- ,and triethanolamines are now available. Combined with fatty acids, they yield soaps which are highly soluble in water.

2. Common Salt or Sodium Chloride

Sodium chloride is a unique raw material in soap manufacture. For good quality soaps salt used must be free from compounds of iron, calcium and magnesium. Salt may be used either in the powder form or as a brine.Salt is used over and over again, being recovered from the soap lyes during their processing for the manufacture of glycerine.

3. Fats, Fatty Acids and Soap Stocks

There are a number of fats and oil but their choice is limited when their soap making properties, availability and prices are considered. Fatty oils may be classified as nut oils, hard fats and soft oils.

(1) Nut oils

There are characterized by a large proportion of low molecular weight fatty acids, especially lauric acid. These include babassu, coconut and palm kernel.

(2) Hard fats

These contain appreciable quantities of palmitic and stearic acids. These include palm mowrah, animal tallows, greases and hydrogenated oils.

(3) Soft oils

The soft fats have substantial proportions of unsaturated acids, viz. oleic, linoleic and linolenics acids. Hydrogenation of soft oils has widened the field of raw materials. These include olive, cottonseed, groundnut, castor, tall oil and marine oils. Groundnut oil, cottonseed oil and marine oils are commonly hydrogenated for use in soap manufacture.

The fats used in soap manufacture vary considerably in composition. Followings are given together with the names and approximate percentages of the principal fatty acids contained in them(Table 6-3). These acids being present in the form of their glyceryl esters.

Table 6-3 The names and approximate percentages of the principal fatty acids

Fat or Oil	Palmitic	Stearic	Oleic	Fatty Acids Linoleic	Myristic	Lauric
palm oil	35	8	48	7	1	—
tallow	20~30	15~30	40~50	0~5	2	—
cottonseed oil	23	—	23	54	—	—
olive oil	7	2	85	6	—	—
caster oil	—	3	9	3	—	—

4. Household Soaps

(1) Filled soaps

These types of soaps are generally produced by the semi-boiled process. Fillers used are sodium carbonate, sodium silicate, clays or sodium sulphate. These fillers may be added to the hot fluid soap in the crutcher.

① Composition of soap filled with mixed fillers

	parts
Caustic soda solution(18%~20%)	20
Soda ash solution(15%)	20
Coconut oil	20
Sodium silicate(diluted with equal quantity of water)	20
Sodium chloride solution	20

② A heavy silicate soap composition

	kgs
Coconut oil	10.5
33% Sodium hydroxide solution	6.25
Sodium silicate solution of 48°Bé strength	193
Groundnut oil	10.5

Soap cutting	10.5

③ A low silicated soap composition

	kgs
Coconut oil	43
Groundnut oil	15
Castor oil	3
Sodium silicate solution of 45°Bé	75
Sodium hydroxide solution of 30°Bé	35
Soap cutting	100

Note: Filled soaps have drawbacks that they shrink, over harden and effloresce on storages. Laundry soap(yellow in color) belongs to this class. This type of soap is heavily rosined, the proportion of rosin being as high as 50% of the total fat charge. Drawback with this soap is that it is highly soluble and thus wasteful.

(2) Dry cleaning soaps

Dry cleaning is a term used to describe the process of cleaning fabrics with certain organic liquids which act as solvents grease. A considerable amount of dirt, which adheres to the fabric by means of the grease, will be removed during the process, but the cleaning action depends essentially on dissolving out the grease. The solvent does not penetrate into the fibers, i.e.the fabric does not become "wetted", as in the ordinary process of washing.

Most of the solvents which are used for drying cleaning are immiscible with water, and fabrics which are to be dry cleaned should first be well dried and then freed as far as possible from loose dirt and dust. The boiling point of the solvent used for dry cleaning purposes should lie preferably between 80℃ and 120℃. Benzene, light petroleum, petrol is most extensively used on large scale. Benzene is sold according to its specific gravity which should be 0.72~0.78. Carbon tetra-chloride, ethylene trichloride or trichloroethylene is also used as noninflammable solvent in dry cleaning industry.

Some of the soaps which have found their outlet as the dry cleaning soaps are soaps made from triethanolamine, for example triethanolamine oleate and soaps made from esterification glycols with fatty acids for example diglycol oleate, laurate and palmitate etc.

Plasticizers

Lesson Four Plasticizers
增塑剂

Plasticizers or dispersants are additives that increase the plasticity or fluidity of the material to which they are added; these include plastics, cement, concrete, wallboard, and clay. Although the same compounds are often used for both plastics and concretes the desired effect is slightly different. The worldwide market for plasticizers in 2004 had a total volume of around 5.5 million tons, which led to a turnover of just over 6 billion pounds. Plasticizers for concrete soften the mix before it hardens, increasing its workability or reducing water, and are usually not intended to affect the properties of the final product after it hardens. Plasticizers for wallboard increase fluidity of the mix, allowing lower use of water and thus reducing energy to dry the board.

The plasticizers for plastics soften the final product increasing its flexibility. Superplasticizers or high range water reducers or dispersants are chemical admixtures that can be added to concrete mixtures to improve workability. Unless the mix is "starved" of water, the strength of concrete is inversely proportional to the amount of water added or water-cement (w/c) ratio. In order to produce stronger concrete, less water is added (without "starving" the mix), which makes the concrete mixture very unworkable and difficult to mix, necessitating the use of plasticizers, water reducers, superplasticizers or dispersants.

Superplasticizers are also often used when pozzolanic ash is added to concrete to improve strength. This method of mix proportioning is especially popular when producing high-strength concrete and fiber-reinforced concrete. Adding 1%-2% superplasticizers per unit weight of cement is usually sufficient. However, note that most commercially available superplasticizers come dissolved in water, so the extra water added has to be accounted for in mix proportioning. Adding an excessive amount of superplasticizers will result in excessive segregation of concrete and is not advisable. Some studies also show that too much superplasticizers will result in a retarding effect.

Plasticizers are commonly manufactured from lignosulfonates, a by-product from the paper industry. High Range Superplasticizers have generally been manufactured from sulfonated naphthalene condensate or sulfonated melamine formaldehyde, although new-generation products based on polycarboxylic ethers are now available. Traditional lignosulfonate-based plasticizers, naphthalene and melamine sulfonate-based superplasticizers disperse the flocculated cement particles through a mechanism of electrostatic repulsion (see colloid). In normal plasticizers, the active substances are adsorbed on to the cement particles, giving them a negative charge, which leads to repulsion between particles. Lignin, naphthalene and melamine sulfonate superplasticizers are organic polymers. The long molecules wrap themselves around the cement particles, giving them a highly negative charge so that they repel each other.

Polycarboxylate ethers (PCE) or just polycarboxylate (PC), the new generation of superplasticizers, are not only chemically different from the older sulfonated melamine and naphthalene-based products, but their action mechanism is also different, giving cement dispersion by steric stabilization, instead of electrostatic repulsion. This form of dispersion is more powerful in its effect and gives improved workability retention to the cementitious mix. Furthermore, the chemical structure of PCE allows for a greater degree of chemical modification than the older-generation products, offering a range of performance that can be tailored to meet specific needs.

In ancient times, the Romans used animal fat, milk and blood as a superplasticizer for their concrete mixes.

Plasticizers can be obtained from a local concrete manufacturer. Household washing up liquid may also be used as a simple plasticizer.

For gypsum wallboard production

Superplasticizers or dispersants, are chemical additives that can be added to wallboard stucco mixtures to improve workability. In order to reduce the energy in drying wallboard, less water is added, which makes the gypsum mixture very unworkable and difficult to mix, necessitating the use of plasticizers, water reducers or dispersants.

Adding about two pounds of dispersant per thousand square feet in 1/2 inch wallboard (15 g/m² of wallboard) is usually sufficient. Some studies also show that too much of lignosulfonate dispersant could result in a set-retarding effect. Data showed that amorphous crystal formations occurred that detracted from the mechanical needle-like crystal interaction in the core, preventing a stronger core. The sugars, chelating agents in lignosulfonates such as aldonic acids and extractive compounds are mainly responsible for set retardation. These low range water reducing dispersants are commonly manufactured from lignosulfonates, a by-product from the paper industry.

High range superplasticizers (dispersants) have generally been manufactured from sulfonated naphthalene condensate, although new generation products based on polycarboxylic ethers are now available. These high range water reducers are used at 1/2 to 1/3 of the lignosulfonate types.

Traditional lignosulfonate and naphthalene sulfonate based superplasticizers disperse the flocculated gypsum particles through a mechanism of electrostatic repulsion (see colloid). In normal plasticizers, the active substances are adsorbed on to the gypsum particles, giving them a negative charge, which leads to repulsion between particles. Lignin and naphthalene sulfonate plasticizers are organic polymers. The long molecules wrap themselves around the gypsum particles, giving them a highly negative charge so that they repel each other.

For plastics

Plasticizers for plastics are additives, most commonly phthalates, that give hard plastics like PVC the desired flexibility and durability. They are often based on esters of polycarboxylic acids with linear or branched aliphatic alcohols of moderate chain length. Plasticizers work by embedding themselves between the chains of polymers, spacing them apart (increasing the "free volume"), and thus significantly lowering the glass transition temperature for the plastic and making it softer. For plastics such as PVC, the more plasticiser added, the lower its cold flex temperature will be. This means that it will be more flexible, though its strength and hardness will decrease as a result of it. Some plasticizers evaporate and tend to concentrate in an enclosed space; the "new car smell" is caused mostly by plasticizers evaporating from the car interior.

Words and Expressions

plasticizer ['plæstə,saɪzə]	n. 增塑剂
cement [sɪ'ment]	n. 水泥
pozzolanic [pɒtsə'lɑːnɪk]	adj. 凝硬性的，火山灰的
electrostatic [ɪ'lektrəʊ'stætɪk]	adj. 静电的，静电学的
adsorb [əd'zɔːrb]	v. 吸附

aldonic acid	糖醛酸
gypsum [ˈdʒɪpsəm]	n. 石膏
lignin [ˈlɪgnɪn]	n. 木质素
naphthalene [ˈnæfθəliːn]	n. 萘
wrap [ræp]	v. 包，裹，卷
repel [rɪˈpel]	vt. 排斥，相斥
phthalate [ˈθæleɪt]	n. 邻苯二甲酸盐
aliphatic [ˌæləˈfætɪk]	adj. 脂肪族的
embed [ɪmˈbed]	vt. 把……嵌入，埋入

Notes

1. Plasticizers or dispersants are additives that increase the plasticity or fluidity of the material to which they are added; these include plastics, cement, concrete, wallboard, and clay.

译文：增塑剂是能提高材料的流动性的一种添加剂，材料包括塑料、水泥、混凝土、墙板和黏土。

2. Plasticizers for plastics are additives, most commonly phthalates, that give hard plastics like PVC the desired flexibility and durability. They are often based on esters of polycarboxylic acids with linear or branched aliphatic alcohols of moderate chain length.

译文：塑料用增塑剂是一种能提高硬塑料如 PVC 流动性和耐用性的增加剂，通常是邻苯二甲酸酯。增塑剂通常是由多元羧酸和中等链长度的直链或支链脂肪醇反应得到的酯。

Reading Comprehension

1. What are the common plasticizers?
2. Write out the mechanism of plasticizers.

Reading Material

Additives for Rubber Formation　橡胶成形添加剂

Accelerators

As a general, synthetic rubbers require more accelerators and less sulphur than does natural rubber. This is because in the polymer chain there is a proportion of non-vulcanisable structure which does not require sulphur, and in some rubbers, for example SBR (styrene butadiene rubber), this reduces the reactivity of the butadiene sections of the chain. There is also an important and noticeable difference between SBR and NR (natural rubber) with regard to the smaller difference of speed between various classes of accelerator. This is what could be described as ultrafast in NR and becomes medium fast in SBR. These points have to be carefully watched when using blends of NR and SBR.

The accelerators should be adequate for the majority operations likely to be encountered in everyday use. However, where exceptions to this are encountered, or if special characteristics are required, then the various accelerator suppliers and manufactures should be consulted. A general compounder cannot be expected to know everything about each particular process he uses, and should seek the advice of specialist in the field concerned. Much valuable time and perhaps ultimate processing difficulties in the factory can be saved by adopting this policy.

The majority of rubber recipes in use, incorporate organic acceleration, but other rubbers especially CR (chloroprene rubber) types and GR (general rubber) types, use inorganic oxides, such as zinc oxide, to achieve a state of vulcanisation. Others such as EPDM (ethylene propylene diene copolymer) occasionally use organic peroxides, but these are the exception rather than the rule. With normal industrial use, these exceptions will soon disappear. In the meantime, it is always advisable to check the literature if specific compounds of rubbers are in request. Indeed even when experienced it is good practice to check often, as errors can creep in on occasion.

Fillers and extenders

In general fillers can be termed reinforcing and non-reinforcing, and also fall into black and non-black types. The black grades can be available currently in bulk. In compounding, they will be selected either singly or blended to meet and give the desired properties required of the polymer and product in question.

Dilution with non-black material, such as china clay, talc and/or whiting is also practised with general rubber, and many industrial goods. This is not only to keep the cost down, which is very important, but also as a device to "smooth" out the compound to help its processability, especially in extending, calendering operations. Practical experience indicates that it is necessary to achieve processability by the use of these non-back materials, at the right cost levels.

If non-back reinforcement is required, then the pure silica type materials are capable of very good physical properties and are widely used for this purpose, especially with the synthetic rubbers. However, it is of interest to know that carbon black as an ingredient is still the outstanding reinforcement material for general specialists say, "you can have any colour you like, provided it is black!"

Other materials which very loosely could be called fillers (although extender is perhaps a better or rather the correct word) are factice and mineral rubber. Both of these materials extend the hydrocarbon content in their various ways, thus helping to kill nerve and assist breakdown of the polymer during mastication, and once again making the compound subsequently much "smoother" to process.

When considering the choice of filler, it is necessary to ensure its suitability in the service environment of the product. As an obvious example, if chemical resistance is required, and in particular resistance to acid conditions, then whiting (calcium carbonate) is unsuitable, whereas inert materials such as silica and silicates (china clay, talc) are ideal. For extreme acid conditions, then barytes (barium sulphate) is more suitable.

Unit Six Fine Chemicals
精细化学品

History, Inheritance and Development
Scientists Find Cheap Way to Turn Gases Into Liquids

Sinopec employees check natural gas equipment

A research group from ShanghaiTech University has developed a novel method of using light to transform methane, ethane and other gases into valuable liquid products. A paper describing the technique - Selective Functionalization of Methane, Ethane, and Higher Alkanes by Cerium Photocatalysis - was published in Science recently. "The finding provides a new, sustainable and mild catalytic platform for natural gas utilization, and will lead to more applications in the energy and chemical industries," said Zuo Zhiwei, one of the authors of the paper. Zuo is an assistant professor at ShanghaiTech University's physical science and technology school.

Methane, the main component of natural gas, is viewed as a clean fuel and inexpensive feedstock in the chemical community. However, given its natural state as a gas, it is difficult to transport economically. David MacMillan, a member of the National Academy of Sciences in the United States, commented that potential uses of the new method in sectors such as pharmaceuticals, agrochemicals and fine chemicals are clearly evident. "This is a remarkable paper that will be widely influential on a global scale," MacMillan said. Alexander van der Made and Sander Van Bavel, scientists at Future Energy Technologies of Shell Global Solutions, said the paper "presents a key first step toward a green route to activate alkanes under mild conditions".

Usually achieved with metal catalysts, current processes rely on rare and expensive metals, such as platinum and palladium, and often require high temperatures to provide energy, Zuo said. "We wondered if we could develop a more affordable and sustainable catalytic platform using light energy and an economical catalyst," Zuo said.

The group tried cerium, a metallic element that accounts for more than half the rare earth in China. Its low cost and unique properties were attractive, although it is not usually regarded as a potential catalyst for organic reactions. After 2,202 trials, the researchers developed an efficient platform for the catalytic conversion of methane and other gaseous alkanes under LED irradiation at room temperature using abundant and inexpensive cerium salts as photocatalysts.

Practice and Training

A Detergent Invention Patent

Invention patent

Unit Seven

Analytical Chemistry

分析化学

Lesson One How to Use Analytical Apparatus
如何使用分析仪器

Buret

A buret (see Fig. 7-1) is used to deliver solution in precisely measured, variable volumes. Burets are used primarily for titration, to deliver one reactant until the precise end point of the reaction is reached.

Using a Buret

To fill a buret, close the stopcock at the bottom and use a funnel. You may need to lift up on the funnel slightly, to allow the solution to flow in freely.

You can also fill a buret using a disposable transfer pipet. This works better than a funnel for the small, 10 ml burets. Be sure the transfer pipet is dry or conditioned with the titrant, so the concentration of solution will not be changed.

Before titrating, condition the buret with titrant solution and check that the buret is flowing freely. To condition a piece of glassware, rinse it so that all surfaces are coated with solution, then drain. Conditioning two or three times will insure that the concentration of titrant is not changed by a stray drop of water.

Check the tip of the buret for an air bubble. To remove an air bubble, whack the side of the buret tip while solution is flowing. If an air bubble is present during a titration, volume readings may be in error. Rinse the tip of the buret with water from a wash bottle and dry it carefully. A minute later, check for solution on the tip to see if your buret is leaking. The tip should be clean and dry before you take an initial volume reading.

When your buret is conditioned and filled, with no air bubbles or leaks, take an initial volume reading. A buret reading card with a black rectangle can help you to take a more accurate reading. Read the bottom of the meniscus. Be sure your eye is at the level of meniscus, not above or below. Reading from an angle, rather than straight on, it results in a parallax error.

Deliver solution to the titration flask by turning the stopcock. The solution should be delivered quickly until a couple of ml from the end point.

The end point should be approached slowly, a drop at a time. Use a wash bottle to rinse the tip of the buret and the sides of the flask.

Pipet

A pipet (see Fig. 7-2) is used to measure small amounts of solution very accurately. A pipet bulb is used to draw solution into the pipet.

Using a Pipet

Start by squeezing the bulb in your preferred hand. Then place the bulb on the flat end of the pipet. Place the tip of the pipet in the solution and release your grip on the bulb to pull solution into the pipet. Draw solution in above the mark on the neck of the pipet. If the volume of the pipet is larger than the volume of the pipet bulb, you may need to remove the bulb from the pipet and squeeze it and replace it on the pipet a second time, to fill the pipet volume completely.

Quickly, remove the pipet bulb and put your index finger on the end of the pipet. Gently release the seal made by your finger until the level of the solution meniscus exactly lines up with the mark on the pipet. Practice this with water until you are able to use the pipet and bulb consistently and accurately.

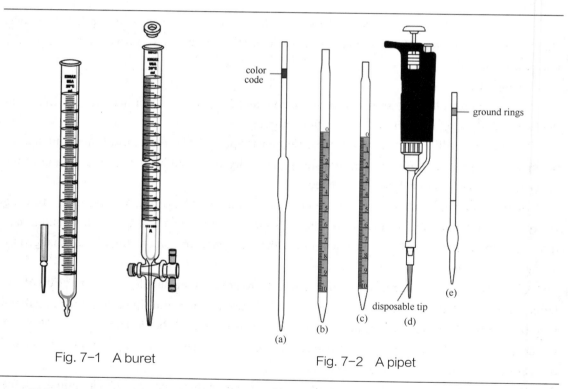

Fig. 7-1 A buret   Fig. 7-2 A pipet

Volumetric Flask

A volumetric flask (see Fig. 7-3) is used to make up a solution of fixed volume very accurately. This volumetric flask measures 500 mL ± 0.2 mL. This is a relative uncertainty of 4×10^{-4} or 400 parts per million.

Using Volumetric Flask

To make up a solution, first dissolve the solid material completely, in less water than required to fill the flask to the mark.

After the solid is completely dissolved, very carefully fill the flask to the 500 mL mark. Move your eyes to the level of the mark on the neck of the flask and line it up so that the circle around the neck looks like a line, not an ellipse. Then add distilled water a drop at a time until the bottom of the meniscus lines up exactly with the mark on the neck of the flask. Take care that no drops of liquid are in the neck of the flask above the mark. After the final dilution, remember to mix your solution thoroughly, by inverting the flask and shaking.

Titration

Begin by preparing your buret, as described on the buret page. Your buret should be conditioned and filled with titrant solution. You should check for air bubbles and leaks, before proceeding with the titration.

Doing a Titration

Take an initial volume reading and record it in your notebook. Before beginning a titration, you should always calculate the expected end point volume.

Prepare the solution to be analyzed by placing it in a clean Erlenmeyer flask or beaker. If your sample is a solid, make sure it is completely dissolved. Put a magnetic stirrer in the flask and add an indicator.

Use the buret to deliver a stream of titrant to within a couple of ml of your expected end point. You will see the indicator change color when the titrant hits the solution in the flask, but the color change disappears upon stirring.

Approach the end point more slowly and watch the color of your flask carefully. Use a wash bottle to rinse the sides of the flask and the tip of the buret, to be sure all titrant is mixed in the flask (see Fig. 7-4)

Fig. 7-3 A volumetric flask

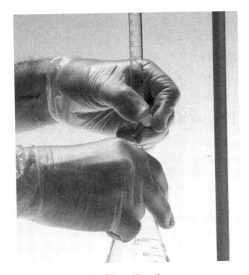

Fig. 7-4 The titration process

As you approach the end point, you may need to add a partial drop of titrant. You can do this with a rapid spin of a Teflon stopcock or by partially opening the stopcock and rinsing the partial drop into the flask with a wash bottle.

Make sure you know what the end point should look like. For phenolphthalein, the end point is the first permanent pale pink. The pale pink fades in 10 to 20 minutes.

If you think you might have reached the end point, you can record the volume reading and add another partial drop. Sometimes it is easier to tell when you have gone past the end point. When you have reached the end point, read the final volume in the buret and record it in your notebook.

Subtract the initial volume to determine the amount of titrant delivered. Use this, the concentration of the titrant, and the stoichiometry of the titration reaction to calculate the number of moles of reactant in your analyte solution.

Advance Bio amino acid analysis

Words and Expressions

buret [bjʊr'ret]	n. 滴定管，玻璃量管
titration ['taɪtreɪʃən]	n. 滴定
reactant [riː'æktənt]	n. 反应物
stopcock ['stɒpkɒk]	n. 管闩，活塞,活栓，旋塞阀
funnel ['fʌnəl]	n. 漏斗；（蒸汽火车或轮船等的）烟囱
	vt. & vi. 倾销
pipet [pɪ'pet]	n. 吸量管，球管
titrant ['taɪtrənt]	n. 滴定剂，滴定（用）标准液
meniscus [mɪ'nɪskəs]	n. 新月，半月板
volumetric flask	（容）量瓶
ellipse [ɪ'lɪps]	n. 椭圆
erlenmeyer flask	锥形烧瓶，爱伦美氏（烧）瓶（锥形平底短颈，广泛用于化学实验室）
beaker ['biːkə]	n. 高脚杯，烧杯
magnetic stirrer	磁搅拌器，磁力搅拌机，磁性搅拌器，电磁搅拌器
indicator ['ɪndɪkeɪtə]	n.（仪器上显示温度、压力、耗油量等的）指针，指示器，记录器；（车辆上的）转弯指示灯，方向灯；指示物，指示者；指示信号，标志，迹象；指示剂
partial ['pɑːʃəl]	adj. 部分的，不完全的；偏爱的；偏袒的
teflon ['teflɑːn]	n. 特氟龙
subtract [səb'trækt]	vt. & vi. 减，扣除，做减法
stoichiometry [ˌstɔɪkɪ'ɒmɪtrɪ]	n. 化学计算（法）；化学计量学
analyte [ænə'lɪt]	n.（被）分析物

Notes

1. After a minute, check for solution on the tip to see if your buret is leaking.

译文：静置片刻，检查滴定管尖端溶液观察滴定管是否漏液。

2. If the volume of the pipet is larger than the volume of the pipet bulb, you may need to remove the bulb from the pipet and squeeze it and replace it on the pipet a second time, to fill the pipet volume completely.

译文：如果移液管体积比洗耳球体积大，将洗耳球从移液管管端移出，挤压洗耳球，将其重新放置到管端，使溶液充满整个移液管。

3. Move your eyes to the level of the mark on the neck of the flask and line it up so that the circle around the neck looks like a line, not an ellipse.

译文：将视线移至容量瓶瓶口的刻度线处并与其保持水平，这样瓶颈处的一周看上去就像是一条线而不是椭圆。

4. If you think you might have reached the end point, you can record the volume reading and add another partial drop. Sometimes it is easier to tell when you have gone past the end point.

译文：如果认为有可能已经达到滴定终点，就将容积的读数记录下来，然后滴加半滴。有时候很容易就能判断出滴定终点是否已滴过。

Reading Comprehension

1. How to use buret and pipet?
2. How to determine the concentration of the analyte?
3. How to prepare standard solutions using volumetric flask?

Reading Material

Unit Operations of Analytical Chemistry 分析化学的单元操作

Measuring Mass with an Electronic Balance

The electronic balance has many advantages over other types of balance. The most obvious is the ease with which a measurement is obtained. All that is needed is to place an object on the balance pan and the measurement can be read on the display to hundredths of a gram. A second advantage, using the Zero button on the front of the balance, is less recognized by students of science beginning. Because one must never place a chemical directly on the balance pan, some container must be used. Place the container on the balance and the mass of the container will be displayed. By pressing the Zero button at this point, the balance will reset to zero and ignore the mass of the container. You may now place the substance to be weighed into the container and the balance will show only the mass of the substance. This saves calculation time and effort. However, when the container is removed from the balance, the display will go into negative numbers until the Zero button is pressed again.

Our electronic balances also have a Unit button on the front. Pressing this button will change the units being measured. Since we have very few times when we need something other than metric units, you should not have to change the mode on the balance. Because of the Unit and Zero buttons, there are two things you must always do before placing objects onto an electric balance to be measured:

See that the display is reading 0.00

See that the unit sign in the upper right of the display shows g

When finished with the electronic balance, press the ON/OFF button and hold it down until the display shows OFF.

Weighing bottles

Solids are conveniently dried and stored in weighing bottles, two common varieties of which are shown in Fig. 7-5. The ground-glass portion of the cap-style bottle shown on the left is on the outside and does not come into contact with the contents; this design eliminates the possibility of some of the sample becoming entrained on and subsequently lost from the ground-glass surface.

Plastic weighing bottles are available; ruggedness is the principal advantage of these bottles over their glass counterparts.

Fig. 7-5 Weighing bottle

Preparation of a Filter Paper

Fig. 7-6 shows the sequence for folding and seating a filter paper in a 60-degfunnel. The paper is folded exactly in half (a), firmly creased, and folded again(b). A triangular piece from one of the corners is torn off parallel to the second fold(c). The paper is then opened so that the untorn quarter forms a cone (d). The cone is fitted into the funnel, and the second fold is creased (e). Seating is completed by dampening the cone with water from a wash bottle and gently patting it with a finger (f). There will be no leakage of air between the funnel and a properly seated cone; in addition, the stem of the funnel will be filled with an unbroken column of liquid.

Filtering a Precipitate from a Solution

Filtering a solid out of a liquid is done using filter paper and a filter funnel. The filter funnel is supported by the ring on a ring stand. Lay a clay triangle across the ring, and then place the filter funnel into the triangle.

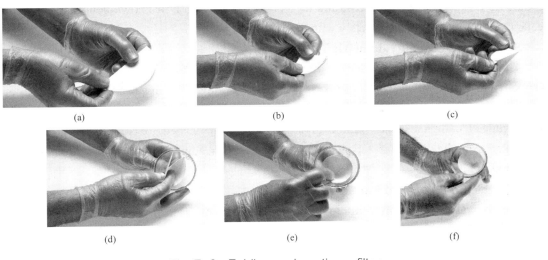

Fig. 7-6 Folding and seating a filter

To prepare the filter paper, fold the paper in half, then fold it in half again. When you look at the open edge of the folded paper you will see four edges of paper. With thumb and finger, catch three of these edges. Squeeze the sides of the folded paper and a con will form with three thicknesses of paper on one side and one thickness of paper on the other. Place this cone of paper into the filter funnel. Place a "catch container" under the stem of the filter funnel and adjust the height of the ring on the ring stand until the tip of the stem is below the mouth of the container.

Use a wash bottle and wet down the inside of the filter paper. This will help it stick to the funnel. You are now ready to filter. Carefully pour the liquid to be filtered into the mouth of the funnel. Do not let the liquid rise to the top of the filter paper. If any liquid goes over and around the paper, your procedure is ruined. Be patient, it will take time for the liquid to move through the pores of the paper. When all the original liquid has been poured into the funnel, use the wash bottle to rinse any remaining precipitate out of the original container. Do not touch or try to stir the liquid inside the filter paper. The wet paper is easily torn, which will ruin your procedure.

If the objective of your filtration is the solid-free liquid, throw the filter paper and its contents into the trash. If your objective is the solid, carefully remove the filter paper and set it in a secure place to dry.

Lesson Two Titrimetric Methods
滴定分析

Titrimetric methods include a large and powerful group of quantitative procedures that are based on measuring the amount of a reagent of known concentration that is consumed by an analyte.

Volumetric titrimetry involves measuring the volume of a solution of known concentration that is needed to react essentially completely with the analyte.

Some Terms Used in Volumetric Titrimetry

A standard solution (or a standard titrant) is a reagent of known concentration that is used to carry out a titrimetric analysis. A titration is performed by slowly adding a standard solution from a buret or other liquid-dispensing device to a solution of the analyte until the reaction between the two is judged complete. The volume or mass of reagent needed to complete the titration is determined from the difference between the initial and final readings. A volumetric titration process is depicted in Fig. 7-7.

Equivalence Points and End Points

The equivalence point in a titration is a theoretical point reached when the amount of added titrant is chemically equivalent to the amount of analyte in the sample. For example, the equivalence point in the titration of sodium chloride with silver nitrate occurs after exactly 1 mol of silver ion has been added for each mole of chloride ion in the sample. The equivalence point in the titration of sulfuric acid with sodium hydroxide is reached after introduction of 2 mol of base for each mole of acid.

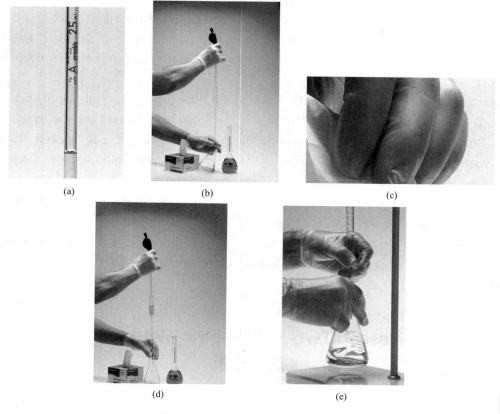

Fig. 7-7 A volumetric titration process

We cannot determine the equivalence point of a titration experimentally. Instead, we can only estimate its position by observing some physical change associated with the condition of equivalence. This change is called the end point for the titration. Every effort is made to ensure that any volume or mass difference between the equivalence point and the end point is small. Such differences do exist, however, as a result of inadequacies in the physical changes and in our ability to observe them. The difference in volume or mass between the equivalence point and the end point is the titration error.

Indicators are often added to the analyte solution to produce an observable physical change (the end point) at or near the equivalence point. Large changes in the relative concentration of analyte or titrant occur in the equivalence-point region. These concentration changes cause the indicator to change in appearance. Typical indicator changes include the appearance or disappearance of a color, a change in color, or the appearance or disappearance of turbidity. As an example, the indicator used in the precipitation titration of silver ion with potassium thiocyanate is a small amount of ferric chloride, which reacts with thiocyanate ions to give a red color. The indicator reaction is

$$Fe^{3+} + SCN^- \longrightarrow [Fe(SCN)]^{2+}$$

We often use instruments to detect end points. These instruments respond to properties of the solution that change in a characteristic way during the titration. Among such instruments are colorimeters, turbidimeters, temperature monitors, refractometers, voltmeters, current meters, and conductivity meters.

Primary Standards

A primary standard is a highly purified compound that serves as a reference material in volumetric and mass titrimetric methods. The accuracy of a method is critically dependent on the properties of this compound. Important requirements for a primary standard are the following:

(1) High purity. Established methods for confirming purity should be available.

(2) Atmospheric stability.

(3) Absence of hydrate water so that the composition of the solid does not change with variations in humidity.

(4) Modest cost.

(5) Reasonable solubility in the titration solution.

(6) Reasonably large molar mass so that the relative error associated with weighing the standard is minimized.

Very few compounds meet or even approach these criteria and only a limited number of primary-standard substances are available commercially. As a consequence, less pure compounds must sometimes be used in place of a primary standard. The purity of such a secondary standard must be established by carefully analysis.

Standard solutions

Standard solutions play a central role in all titrimetric methods of analysis. Therefore, we need to consider the desirable properties for such solutions, how they are prepared, and how their concentrations are expressed. The ideal standard solution for a titrimetric method will

(1) Be sufficiently stable so that it is necessary to determine its concentration only once.

(2) React rapidly with the analyte so that the time required between additions of reagent is minimized.

(3) React more or less completely with the analyte so that satisfactory end points are realized.

(4) Undergo a selective reaction with the analyte that can be described by a balanced equation.

Few reagents meet all these ideals perfectly.

The accuracy of a titrimetric method can be no better than the accuracy of the concentration of the standard solution used in the titration. Two basic methods are used to establish the concentration of such solutions. The first is the direct method, in which a carefully weighed quantity of a primary standard is dissolved in a suitable solvent and diluted to an exactly known volume in a volumetric flask. The second is by standardization, in which the titrant to be standardized is used to titrate (1) a weighed quantity of a quantity of a primary standard, (2) a weighed quantity of a secondary standard, or (3) a measured volume of another standard solution. A titrant that is standardized against a secondary standard or against another standard solution is sometimes referred to as a secondary-standard solution. The concentration of a second-standard solution is subject to a larger uncertainty than is that of a primary-standard solution. If there is a choice, then, solutions are best prepared by the direct method. Many reagents lack the properties required for a primary standard, however, and therefore require standardization.

Words and Expressions

titrimetric [ˌtaɪtrəˈmetrɪk]	*adj.* 滴定（测量）的
quantitative [ˈkwɒntɪtətɪv]	*adj.* 数量（上）的，量化的，定量性的
reagent [riːˈeɪdʒənt]	*n.* 反应物，试剂
standard solution	标准溶液
depict [dɪˈpɪkt]	*vt.* 描绘，描画，描写，描述
equivalence point	等当量点
end point	终点
sodium chloride	氯化钠
silver nitrate	硝酸银
sulfuric acid	硫酸
sodium hydroxide	氢氧化钠
titration error	滴定误差
turbidity [ˈtɜːbɪdɪtɪ]	*n.* 混浊，混乱
potassium thiocyanate	硫氰酸钾
ferric [ˈferɪk]	*adj.* 铁的，含铁的，（正）铁的，三价铁的
turbidimeter [ˌtəbɪˈdɪmɪtə]	*n.* 浊度计，浊度表
refractometer [ˌriːfrækˈtɒmɪtə]	*n.* 折射计
primary-standard	基准物
standardization [ˌstændədaɪˈzeɪʃən]	*n.* 标准化，标定

Notes

1. Titrimetric methods include a large and powerful group of quantitative procedures that are based on measuring the amount of a reagent of known concentration that is consumed by an analyte.

译文：滴定分析包含很多定量分析方法，它是测定被分析物所消耗的已知浓度试剂的量的分析方法。

2. The equivalence point in a titration is a theoretical point reached when the amount of added titrant is chemically equivalent to the amount of analyte in the sample.

译文：滴定中的化学计量点是当所加入的滴定剂的量按化学计量关系等于样品中被分析的量时所达到的理论终点。

3. The difference in volume or mass between the equivalence point and the end point is the titration error.

译文：化学计量点和滴定终点之间的体积或质量的差异称为滴定误差。

4. The first is the direct method, in which a carefully weighed quantity of a primary standard is dissolved in a suitable solvent and diluted to an exactly known volume in a volumetric flask.

译文：第一种方法是直接法，用适宜的溶剂溶解精确称量的基准物质，将其在容量瓶中稀释到精确已知的体积。

Reading Comprehension

1. Distinguish between (a) the equivalence point and the end point of a titration; (b) a primary standard and a secondary standard.

2. What solution can be a standard solution used in a titrimetric analysis?

3. What is indicator change?

Reading Material

Systematic Errors　系统误差

Systematic errors have a definite value and an assignable cause, and are of the same magnitude for replicate measurements made in the same way. Systematic errors lead to bias in measurement results. Note that bias affects all of the data in a set in the same way and that it bears a sign.

Sources of Systematic Errors

There are three types of systematic errors: (1) Instrumental errors are caused by nonideal instrument behavior, by faulty calibrations, or by use under inappropriate of analytical systems. (2) Method errors arise from nonideal chemical or physical behavior of analytical systems. (3) Personal errors result from the carelessness, inattention, or personal limitations of the experimenter.

Instrument Errors

All measuring devices are potential sources of systematic errors. For example, pipets, burets, and volumetric flasks may hold or deliver volumes slightly different from those indicated by their graduations. These differences arise from using glassware at a temperature that differs significantly from the calibration temperature, from distortions in container walls due to heating while drying, from errors in the original calibration, or from contaminants on the inner surfaces of the containers. Calibration eliminates most systematic errors of this type.

Method Errors

The nonideal chemical or physical behavior of the reagents and reactions on which an analysis is based often introduces systematic method errors. Such sources of nonideality include the slowness of some reactions, the incompleteness of others, the instability of some species, the nonspecificity of most reagents, and the possible occurrence of side reactions that interfere with the measurement process. For example, a common method error in volumetric analysis results from the small excess of reagent required to cause an indicator to undergo the color changes that signals completion of the reaction. The accuracy of such an analysis is thus limited by the very phenomenon that makes the titration possible.

Error inherent in a method are often difficult to detect and thus the most serious of the three types of systematic error.

Personal Errors

Many measurements require personal judgments. Examples include estimating the position of a pointer between two scale divisions, the color of a solution at the end point in a titration, or the level of a liquid with respect to a graduation in a pipet or buret. Judgements of this type are often subject to systematic, unidirectional errors. For example, one person may read a pointer consistently high, another may be slightly slow in activating a timer, and a third may be less sensitive to color changes. An analyst who is insensitive to color changes tends to use excess reagent in a volumetric analysis. Analytical procedures should always be adjusted so that any known physical limitations of the analyst cause negligibly small errors.

A universal source of personal error is prejudice, or bias. Most of us, no matter how honest, have a natural tendency to estimate scale readings in a direction that improves the precision in a set of results. Alternatively, we may have a preconceived notion of the true value for the measurement. We then subconsciously cause the results to fall close to this value. Number bias is another source of personal error that varies considerably from person to person. The most frequent number bias encountered in estimating the position of a needle on a scale involves a preference for the digits 0 and 5. Also common is a prejudice favoring small digits over large and even number over odd.

Lesson Three   Gas Chromatography
气相色谱

Description

Chromatography is the science of separation, which uses a diverse group of methods to separate closely related components of complex mixtures. During gas chromatographic separation, the sample is transported via an inert gas called the mobile phase. The mobile phase carries the sample through a coiled tubular column where analytes interact with a material called the stationary phase. For separation to occur, the stationary phase must have an affinity for the analytes in the sample mixture. The mobile phase, in contrast with the stationary phase, is inert and does not interact chemically with the analytes. The only function of the mobile phase is to sweep the analyte mixture through the length of the column. Gas chromatography can be divided into two categories: (1) gas-solid and (2) gas-liquid chromatography. Gas-liquid GC, developed in 1941, is the primary GC technique used for environmental applications. Gas-solid GC is not widely used for environmental applications.

The stationary phase is chosen so that the components of the sample distribute themselves between the mobile and stationary phase to varying degrees. Those components that are strongly retained by the stationary phase move slowly relative to the flow of the mobile phase. In contrast, components that have a lower affinity for the stationary phase travel through the column at a faster rate. As consequence of the differences in mobility, sample components separate into discrete bands that can be analyzed qualitatively and quantitatively.

Gas chromatography is the most widely used chromatographic technique for environmental analyses.

System Components

The primary components of a GC include:

Injection port

Column

Integrator or data acquisition system

Detectors

Other parts include:

Autosampler

Control panel, electronic pressure control (EPC)

Injection port liners

Septa

Ferrules

Flow controllers

The carrier gas is introduced in the injection port where the sample is volatilized and swept through the column, and where the compounds are separated. The carrier gas/sample mixture then enters the detector where the compounds are identified. The signal from the detector then is amplified and displayed by the data system.

A capillary column is an open tube made of fused silica with an outer coating of durable plastic and an inner coating of stationary-phase material. Some capillary columns have a second outer covering of stainless steel to withstand the higher pressure required to analyze alcohols, ketone, and VOCs by the purge-and-trap method. A lesser used column type is the packed column. Packed columns use a stainless steel or glass tube with a 1/8th inch inner diameter packed with a solid stationary phase.

The effectiveness of a chromatographic column in separating solutes (analytes) is dependent on a number of variables. Understanding these variables is essential to the process of optimizing any chromatographic system and achieving resolution of analytes. Variables that affect separation include distribution equilibrium constants, retention time, retention (capacity) factors, and selectivity factors.

Theory of Operation

The theory of separation by GC is relatively simple and understanding the factors that affect separation allows more effective applications of GC analysis in the field. The purpose of separation is to allow identification and quantitation of individual components of a mixture and the theory of separation is detailed below. In addition to separation, detection of analytes after separation, which is an essential but separate aspect of chromatography, is presented in the section describing system components. Basic components of a complete gas chromatographic system include:

(1) A carrier gas supply,

(2) A syringe for sample introduction,

(3) The injection port,

(4) The column and oven, and

(5) The detector and data collection system.

Components of a gas chromatograph are presented in greater detail in the section describing the system components. Schematic diagrams and photographs of instruments can also be accessed through the systems components section.

Before separation occurs in the chromatographic column, the mixture of components in the sample is introduced into the chromatograph through the injection port with a syringe. At this point, the analytes are vaporized (if not already in the gas phase) by the high temperature maintained in the injection port. The analytes are kept in the gaseous state by maintaining all elements of the instrument at a temperature above the boiling point of the analytes. The gas phase analytes are then immediately swept onto the chromatographic column by the mobile phase. The mobile phase is comprised of an inert carrier gas, which usually is nitrogen, helium, or hydrogen.

As the analytes are swept through the column by the mobile phase, separation occurs based on the affinity of each analyte for the stationary phase. The gas chromatographic column is composed of a coiled, tubular column and the stationary phase within the tube. GC columns are either packed or open-tubular. Early GC columns were packed with carbon or diatomaceous earth based solids which acted as the stationary phase. In modern open-tubular column, the stationary phase is a liquid organic compound that is coated on the internal surface of the fused silica column. Polarities of the analytes dictate the choice of stationary phase. Components of the mixture with a high degree of affinity for the stationary phase migrate rapidly through the column. As a consequence of the differences in mobility due to affinities for the stationary phase, sample components separate into discrete band that can be qualitatively and quantitatively analyzed.

As individual components of the mixture elute the chromatographic column, they are swept by the carrier gas to a detector. The detector generates a measurable electrical signal, referred to as peaks, that is proportional to the amount of analyte present. Detector response is plotted as a function of the time required for the analyte to elute from the column after injection. The resulting plot is called a chromatogram. Detector response is generally a Gaussian shaped curve representative of the concentration distribution of the analyte band as it elutes from the column. The position of the peaks on the time axis may serve to identify the components and the area under the peaks provide a quantitative measure of the amount of each component.

Words and Expressions

chromatography [ˌkroumə'tɑ:grəfi]	n. 色谱，套色版
inert gas	惰性气体
tubular ['tju:bjulə]	adj. 管状的
control panel	控制面板
amplify ['æmplɪfaɪ]	vt. 放大，增强；v. 扩大
capillary [kə'pɪlərɪ]	n. 毛细管；adj. 毛状的，毛细作用
fused silica	熔融石英
ketone ['ki:təun]	n. [化]酮
mobile phase	流动相
stationary phase	固定相
carrier gas	载气
boiling point	沸点
retention time	保留时间
retention factor	保留因子
selectivity factor	选择性因子
the effectiveness of a chromatographic column	柱效
helium ['hi:ljəm, 'hi:lɪəm]	n. 氦
elute [ɪ'lju:t]	vt. [化]洗提

Notes

1. Chromatography is the science of separation which users a diverse group of methods to separate closely related components of complex mixtures.

译文：色谱法是一种分离技术，该分离技术是使用各种方法来使复杂的混合试样中化学性质相似的组分分离开来。

2. Gas chromatography can be divided into two categories: (1) gas-solid and (2) gas-liquid chromatography. Gas-liquid GC, developed in 1941, is the primary GC technique used for environmental applications. Gas-solid GC is not widely used for environmental applications.

译文：气相色谱法可划分为两大类：(1)气固色谱法和 (2) 气液色谱法。气液色谱法是在1941年发展起来的，它主要用于环保方面。气固色谱法在环保方面应用并不广泛。

3. Those components that are strongly retained by the stationary phase move slowly relative to the flow of the mobile phase. In contrast, components that have a lower affinity for the stationary phase travel through the column at a faster rate.

译文：相对于流动相而言，那些被固定相强有力地保留下来的组分的流速较慢。与此相反，那些和固定相作用不大的组分流经色谱柱时流速较快。

4. Variables that affect separation include distribution equilibrium constants, retention time, retention (capacity) factors, and selectivity factors.

译文：影响分离的变量有分配系数、保留时间、容量因子和选择因子。

Reading Comprehension

1. Speak out components of a GC.
2. How many categories of GC can be divided?
3. How to determine substances using GC and its theory?

Reading Material

Ion Chromatography　离子色谱

Ion-exchange chromatography (or ion chromatography) is a process that allows the separation of ions and polar molecules based on their affinity to the ion exchanger. It can be used for almost any kind of charged molecule including large proteins, small nucleotides and amino acids. The solution to be injected is usually called a sample, and the individually separated components are called analytes. It is often used in protein purification, water analysis, and quality control. Ion-exchange chromatography retains analyte molecules on the column based on coulombic (ionic) interactions. The stationary phase surface displays ionic functional groups (R-X) that interact with analyte ions of opposite charge. This type of chromatography is further subdivided into cation exchange chromatography and anion-exchange chromatography. The ionic compound consisting of the cationic species M^+ and the anionic species B^- can be retained by the stationary phase. Cation exchange chromatography retains positively charged cations because the stationary phase displays a negatively charged functional group:

$$R\text{-}X^-C^+ + M^+B^- \rightleftharpoons R\text{-}X^-M^+ + C^+ + B^-$$

Anion exchange chromatography retains anions using positively charged functional group:

$$R\text{-}X^+A^- + M^+B^- \rightleftharpoons R\text{-}X^+B^- + M^+ + A^-$$

Note that the ion strength of either C^+ or A^- in the mobile phase can be adjusted to shift the equilibrium position, thus retention time. The ion chromatogram shows a typical chromatogram obtained with an anion exchange column.

Typical technique

A sample is introduced, either manually or with an autosampler, into a sample loop of known volume. A buffered aqueous solution known as the mobile phase carries the sample from the loop onto a column that contains some form of stationary phase material. This is typically a resin or gel matrix consisting of agarose or cellulose beads with covalently bonded charged functional groups. The target analytes (anions or cations) are retained on the stationary phase but can be eluted by increasing the concentration of a similarly charged species that will displace the analyte ions from the stationary phase. For example, in cation exchange chromatography, the positively charged analyte could be displaced by the addition of positively charged sodium ions. The analytes of interest must then be detected by some means, typically by conductivity or UV/Visible light absorbance.

In order to control an IC system, a chromatography data system (CDS) is usually needed. In addition to IC systems, some of these CDSs can also control gas chromatography (GC) and HPLC.

Separating proteins

Proteins have numerous functional groups that can have both positive and negative charges. Ion exchange chromatography separates proteins according to their net charge, which is dependent on the composition of the mobile phase. By adjusting the pH or the ionic concentration of the mobile phase, various protein molecules can be separated. For example, if a protein has a net positive charge at pH 7, then it will bind to a column of negatively charged beads, whereas a negatively charged protein would not. By changing the pH so that the net charge on the protein is negative, it too will be eluted.

Elution by increasing the ionic strength of the mobile phase is a more subtle effect—it works as ions from the mobile phase will interact with the immobilized ions in preference over those on the stationary phase. This "shields" the stationary phase from the protein, (and vice versa) and allows the protein to elute.

Separation can be achieved based on the natural isoelectric point of the protein. Alternatively a peptide tag can be genetically added to the protein to give the protein an isoelectric point away from most natural proteins (e.g. 6 arginines for binding to cation-exchange resin such as DEAE-Sepharose or 6 glutamates for binding to anion-exchange resin).

Elution from ion-exchange columns can be sensitive to changes of a single charge-chromatofocusing. Ion-exchange chromatography is also useful in the isolation of specific multimeric protein assemblies, allowing purification of specific complexes according to both the number and the position of charged peptide tags.

History, Inheritance and Development

History of Analytical chemistry

Analytical chemistry has been important since the early days of chemistry, providing methods for determining which elements and chemicals are present in the object in question. During this period significant contributions to analytical chemistry include the development of systematic elemental analysis by Justus von Liebig and systematized organic analysis based on the specific reactions of functional groups.

The first instrumental analysis was flame emissive spectrometry developed by Robert Bunsen and Gustav Kirchhoff who discovered rubidium (Rb) and caesium (Cs) in 1860.

Most of the major developments in analytical chemistry take place after 1900. During this period instrumental analysis becomes progressively dominant in the field. In particular many of the basic spectroscopic and spectrometric techniques were discovered in the early 20th century and refined in the late 20th century.

The separation sciences follow a similar time line of development and also become increasingly transformed into high performance instruments. In the 1970s many of these techniques began to be used together as hybrid techniques to achieve a complete characterization of samples.

Starting in approximately the 1970s into the present day analytical chemistry has progressively become more inclusive of biological questions (bioanalytical chemistry), whereas it had previously been largely focused on inorganic or small organic molecules. Lasers have been increasingly used in chemistry as probes and even to initiate and influence a wide variety of reactions. The late 20th century also saw an expansion of the application of analytical chemistry from somewhat academic chemical questions to forensic, environmental, industrial and medical questions, such as in histology.

Practice and Training

Balance Weighing Exercise

1. Purpose requirements

Understand the structure of analytical balance, learn to use direct weighing method and weight reduction method weighing.

2. Experimental materials

(1) Instrument: watch glass, weighing bottle (NaCl, pallet balance, analytical balance .

(2) Reagent: NaCl

3. Experimental steps

(1) Take the next flat cover, fold, put the flat top right front.

(2) Adjust Balance Level

(3) Inspection

(4) Clean Balance

(5) Adjust Balance Zero

(6) Deweighting NaCl solids (0.2~0.4 g)

Balance weighing exercise

4. Data Record and calculations

No.	Before dumping $m_{NaCl+bottle}$/g	After dumping $m_{NaCl+bottle}$/g	m_{NaCl}/g
1			
2			
3			

Unit Eight
Biochemical Engineering
生物化工

Lesson One Introduction to Biochemical Engineering
生物化工简介

Biochemical engineering addresses a broad range of critical problems in the manufacturing of metabolites, pharmaceuticals, therapeutic proteins and other applications of microbial and animal cell biotechnology. Biochemical engineering provides students with skills to apply chemical engineering principles to bioreactor design, downstream processing and overall bioprocess optimization. The aim is to provide chemical engineers with a basic understanding of the fundamental process engineering problems specific to biochemical processes. Also to provide an insight into the creativity required in bioreactor design.

Research activities include projects extending the Chemical Engineering paradigm of Transport and Kinetics to biological system, as well as projects exploring developments in the Life Science for purpose of providing biological solutions to long-standing engineering problems. The overall objectives are to develop a fundamental understanding of the physical and chemical/biological processes involved in the biosynthesis and transport of a product molecule, and to apply this knowledge to the improvement of the corresponding production and separation/purification operations.

There are projects in genetic and metabolic engineering; new protein expression systems; protein trafficing, post-translational modification and secretion; in-vivo and in-vitro protein folding; interplay between environment and cellular physiology and function; bioreactor design, operations and control; intracellular and extracellular monitoring methods; biocatalysis and bioreactors; membrane and chromatographic operations; reverse micellar and two-phase separation systems; protein-protein and protein-surface interaction; and modelling; expert systems and artificial intelligence applications to problems of upstream and downstream biotechnological operations.

The interdisciplinary nature of biochemical engineering is probably best reflected in the program on Metabolic Engineering. Metabolic engineering is defined as the targeted improvement of cellular properties and function through the modification of a specific biochemical reaction with the use of recombinant DNA methods. Rational pathway modification effected through metabolic engineering has far reaching applications in the optimization of biologically-based production routes, the synthesis of novel materials and drugs, degradation of environmental pollutants, and the medical field for the rational design of therapies through targeted gene modification and supplementation of nutrients.

In a word, biochemical engineering, as an important chemical engineering technique rising in the middle of twenty century, is an interdiscipline between high efficient chemistry & chemical engineering technology and life science, and has many advantages on the basis of interdiscipline. The developing of this discipline must lead a revolutionary change in the field of modern chemistry and chemical engineering.

Words and Expressions

biochemical engineering	生物化工，生化工程
metabolite [me'tæbə‚laɪt]	n. 代谢产物
pharmaceutical [ˌfɑːrməˈsuːtɪkl]	n. 药剂，药物
therapeutic [θerəˈpjuːtɪk]	adj. 治疗的
application [ˌæplɪˈkeɪʃən]	n. 应用
microbial [maɪˈkroʊbɪrl]	adj. 微生物的
biotechnology [ˌbaɪəʊtekˈnɒlədʒɪ]	n. 生物技术
chemical engineering principle	化工原理
paradigm [ˈpærədaɪm, ˈpærədɪm]	n. 范例
bioreactor [baɪəriːˈæktə]	n. 生物反应器
downstream [ˈdaʊnstriːm]	adj. 下游的
bioprocess [baɪəprəˈses]	n. 生物工艺
optimization [ˌɒptɪmaɪˈzeɪʃən]	n. 优化
fundamental [ˌfʌndəˈmentl]	adj. 基本的
kinetics [kaɪˈnetɪks]	n. 动力学
biosynthesis [baɪəˈsɪnθɪsɪs]	n. 生物合成
product [ˈprɒdəkt]	n. 产物
separation [sepəˈreɪʃən]	n. 分离
purification [ˌpjʊərɪfɪˈkeɪʃən]	n. 纯化
operation [ˌɒpəˈreɪʃən]	n. 操作
genetic [dʒɪˈnetɪk]	adj. 遗传的
metabolic [ˌmetəˈbɒlɪk]	adj. 代谢的
expression system	蛋白质表达系统
post-translational modification	翻译后修饰
secretion [sɪˈkriːʃən]	n. 分泌
in vivo	在活的有机体内
in vitro	在体外
folding [ˈfəʊldɪŋ]	n. 折叠
interplay [ˈɪntɜːˈpleɪ]	v. 相互作用

cellular [ˈseljʊlə]	adj.	细胞的
physiology [ˌfɪzɪˈɒlədʒɪ]	n.	生理
intracellular [ˌɪntrəˈseljʊlə]	adj.	细胞内的
extracellular [ˌekstrəˈseljʊlə]	adj.	细胞外的
monitor [ˈmɒnɪtə]	v.	监控
biocatalysis [baɪəkəˈtælɪsɪs]	n.	生物催化
membrane [ˈmembren]	n.	膜
chromatographic [krouˌmætəˈræfɪk]	adj.	色谱的
reverse micellar		反胶团
interaction [ˌɪntərˈækʃen]	n.	相互作用
model [ˈmɒdl]	v.	建模
artificial intelligence		人工智能
application [ˌæplɪˈkeʃen]	n.	应用
upstream [ˈʌpˈstriːm]	adj.	上游的
interdisciplina [ɪntəˈdɪsɪplɪn]	adj.	多学科交叉的
modification [ˌmɒdɪfɪˈkeɪʃen]	n.	修饰，改进
rational [ˈræʃenl]	adj.	合理的
recombinant [riːˈkɒmbənent]	adj.	重组的
degradation [ˌdegrəˈdeɪʃen]	n.	降解

Notes

1. Biochemical engineering addresses a broad range of critical problems in the manufacturing of metabolites, pharmaceuticals, therapeutic proteins and other applications of microbial and animal cell biotechnology.

句子分析：本句中 address 翻译为处理、应对。英文中介词短语作定语一般放在被修饰词的后面，而中文则在被修饰词的前面，翻译时要注意两种语言的差异。

译文：生物化工处理代谢产物、药物、治疗性蛋白的生产中以及微生物与动物细胞生物技术的其它应用中广泛的重要问题。

2. Research activities include projects extending the Chemical Engineering paradigm of Transport and Kinetics to biological system, as well as projects exploring developments in the Life Science for purpose of providing biological solutions to long-standing engineering problems.

句子分析：本句中，Research activities 前省略"生物化工"，翻译时补足；Extending 与 exploring 都是现在分词短语作为 projects 的后置定语的；英文长句在翻译时，可以将之分为若干个中文短句，以免句子结构过于冗长复杂。

译文：（生物化工）研究活动包括将化工原理中的传质与动力学的范例拓展到生物系统的项目，以及探索生命科学的进展的项目，其目的是为长期存在的工程问题提供生物学解决方案。

3. Rational pathway modification effected through metabolic engineering has far reaching applications in the optimizaion of biologically-based production routes, the synthesis of novel materials and drugs, degradation of environmental pollutants, and the medical field for the rational

design of therapies through targeted gene modification and supplementation of nutrients.

句子分析：本句的翻译调整了语序，若按照原文句序翻译，则后面的状语成分过于长，显得结构不匀称。"effect" 此处为动词，用过去分词的形式作后置定语修饰 "rational pathway modification"，意为"实现"。

译文：在以生物学为基础的生产途径优化，独特的物质与药物的合成，环境污染物的降解方面，以及医学领域中通过基因修饰与营养补充进行的治疗方案合理设计方面通过代谢工程加以实现的合理代谢途径修饰目前已经得到了应用。

4. In a word, biochemical engineering, as an important chemical engineering technique rising in the middle of twenty century, is an interdiscipline between high efficient chemistry & chemical engineering technology and life science, and has many advantages on the basis of interdiscipline.

句子分析：本句中 "rising" 现在分词短语作后置定语，"on the basis of" 是 "在……的基础上，基于" 的意思，也可用 "based on" 代替。

译文：总而言之，作为 20 世纪中期兴起的一种重要的化工技术，生物化工是高效的化学与化工技术和生命科学之间的交叉学科，基于其交叉学科的特点，具有很多的优势。

Reading Comprehension

1. What is biochemical engineering?
2. What research activities does biochemical engineering include?
3. Which project can best reflect the interdisciplinary nature of biochemical engineering? How?

Reading Material

Biomass energy

Plants and Organic Waste Offer Hopes of Filling Energy Gap
植物与有机废物有希望弥补能量的不足

Biomass, a source of energy as old as a caveman's fire, may never fully replace the fossil fuels that two centuries of industrial society have nearly exhausted. But some experts believe that biomass, coupled with other energy sources, may at least tide the world over the energy famine that threatens the coming decades.

The word biomass describes all solid materials of animal or vegetable origin from which energy may be extracted. It included products as valuable as cotton fibre and leather, and by-products as nominally as tree branches, peanut shells, spoiled wheat cattails, corn husks, seaweed, garbage, waste paper and cow dung. Biomass can be burned, fermented or reacted chemically with other materials to release energy.

In addition to waste biomass, scientists are experimenting with "energy agriculture"-the cultivation of plants for energy rather than food. Such plants include fermentable cereals from which fuel alcohol can be produced, desert bushes with sap consisting of hydrocarbon oil and even a Brazilian tree that yields diesel oil of such purity it is said to be directly usable by cars.

In whatever form it takes, biomass is an orderly chemical structure built from a chaos of simpler chemical substances using the energy of the sun through the natural process of photosynthesis. By breaking up biomass into its original chaotic state, a part of the solar energy can be extracted in usable form.

In other parts of the world, particularly in China and India, bio-gas, consisting mostly of methane generated by the fermentation of cattle and human feces, is already a major alternative source of fuel.

It is reported recently that some seven million manure-based bio-gas generators were in operation in Sichuan Province, alone, most of them providing cooking gas for small groups of families or local industries.

India's development of bio-gas has been slower but nevertheless impressive. Under its current five-year plan, the government hopes to complete 500,000 bio-gas plants. If all these plants could be completed by 2010, bio-gas would provide the country with the equivalent of 44 percent of its projected electricity needs.

As long as we have had cheap coal, oil and gas in the ground, we could afford to waste the biological fuels all around us. But those days are gone, and we no longer can afford to waste anything.

Words and Expressions

biomass [baɪəmæs]	n. 生物量
caveman ['keɪvˌmæn]	n. （石器时代的）穴居人
fossil fuel	化石燃料
famine ['fæmɪn]	n. 饥荒，严重不足
fibre ['faɪbə]	n. 纤维
leather ['leðə]	n. 皮革
cattail ['kætˌteɪl]	n. 香蒲属植物
corn husk	玉米的壳
seaweed ['siːwiːd]	n. 海草，海藻
cow dung	牛屎
cultivation [ˌkʌltɪ'veɪʃən]	n. 耕作，栽培
alcohol ['ælkəhɒl]	n. 酒精
sap [sæp]	n. 树液
photosynthesis [ˌfəʊtəʊ'sɪnθəsɪs]	n. 光合作用
methane ['meθeɪn]	n. 甲烷
cattle ['kætl]	n. 牛，牲口，家畜
feces ['fiːsiːz]	n. 粪
manure-based	以粪肥为主的
cooking gas	适于烧煮的气

Lesson Two Molecular Structure of Nucleic Acid—A Structure for Deoxyribose Nucleic Acid
核酸的分子结构——脱氧核糖核酸的结构

DNA, chromosomes and genomes

We wish to suggest a structure for the salt of deoxyribose nucleic acid (DNA). This structure has novel features which are of considerable biological interest.

This structure has two helical chains each coiled round the same axis. We have made the usual chemical assumptions, namely, that each chain consists of phosphate diester groups joining β-D-deoxyribofuranose residues with 3', 5'linkages. The two chains (but not their bases) are related by a dyad perpendicular to the fibre axis. Both chains follow right-handed helices, but owing to the dyad the sequences of the atoms in the chains run in opposite directions. The bases are on the inside of the helix and the phosphate on the outsides. The sugar is roughly perpendicular to the attached base. There is a residue on each chain every 3.4 Å in the z-direction. We have assume an angle of 36°between adjacent residues in the same chain, so that the structure repeats after 10 residues on each chain, that is after 34 Å. distance of a phosphorus atom from the fibre axis is 10 Å. As the phosphates are on the outside, cations have access to them.

The structure is an open one, and its water content is rather high. At lower water content we would expect the bases to tilt so that the structure could become more compact.

The novel feature of the structure is the manner in which the two chains are held together by the purine and pyrimidine bases. The planes of the bases are perpendicular to the fibre axis. They are joined together in pairs, a single base from one chain being hydrogen-bonded to a single base from the other chain, so that the two lie side by side with identical z-coordinates. One of the pair must be a purine and the other a pyrimidine for bonding to occur. The hydrogen are made as follows: purine position 1 to pyrimidine position 1; purine position 6 to pyrimidine position 6.

If it is assumed that the bases only occur in the structure in the most plausible tautomeric forms (that is, with the keto rather than the enol configurations) it is found that only specific pairs of bases can bond together. These pairs are: adenine (purine) with thymine (pyrimidine), and guanine (purine) with cytosine (pyrimidine).

In other words, if an adenine forms one member of a pair, on either chain, then on these assumptions the other member must be thymine; similarly for guanine and cytosine. The sequence of bases on a single chain does not appear to be restricted in any way. However, if only specific pairs of bases can be formed, it follows that if sequence of bases on one chain is given, then the sequence on the other chain is automatically determined.

It has been found experimentally that the ratio of the amounts of adenine to thymine, and the ratio of guanine to cytosine, are always very close to unity for deoxyribose nucleic acid.

It is probably impossible to build this structure with a ribose sugar in place of the deoxyribose, as the extra oxygen atom would make too close a van der Waals contact.

It has not escaped our notice that the specific pairing we have postulated immediately suggests a possible copying mechanism for the genetic material.

Words and Expressions

deoxyribose nucleic acid	脱氧核糖核酸
feature ['fi:tʃə]	n. 性质
helical ['helɪkəl]	adj. 螺旋的
coil [kɔɪl]	v. 盘绕
axis ['æksɪs]	n. 轴
assumption [ə'sʌmpʃən]	n. 假设
phosphate diester group	磷酸二酯基团
deoxyribofuranose [di:ɒksɪrɪbɔ:frɒ:'nouz]	n. 脱氧核糖呋喃糖
residue ['rezɪdju:]	n. 残基
linkage ['lɪŋkɪdʒ]	n. 连接
base [beɪs]	n. 碱基
dyad ['daɪˌæd, -əd]	n. 二联体
perpendicular [ˌpɜ:pən'dɪkjʊlə]	adj. 垂直的
adjacent [ə'dʒeɪsənt]	adj. 临近的
phosphorus ['fɒsfərəs]	n. 磷
cation ['kætaɪən]	n. 阳离子
content [kən'tent]	n. 含量
tilt [tɪlt]	v. 倾斜
compact ['kɒmpækt, kəm'pækt]	adj. 紧凑的
purine ['pjʊrɪn]	n. 嘌呤
pyrimidine [paɪ'rɪməˌdi:n]	n. 嘧啶
plane [pleɪn]	n. 平面
z-coordinate	z 轴坐标
position [pə'zɪʃən]	n. 位置
plausible ['plɔ:zəbl]	adj. 稳定的
tautomeric [tɔ:tə'merɪk]	adj. 互变异构的
keto ['ki:tou]	n. 酮式
enol ['i:noʊl]	n. 烯醇式
configuration [kənˌfɪgjʊ'reɪʃən]	n. 构象
adenine ['ædəni:n]	n. 腺嘌呤
thymine ['θaɪmi:n]	n. 胸腺嘧啶
guanine ['gwɑ:ni:n]	n. 鸟嘌呤
cytosine ['saɪtəsi:n]	n. 胞嘧啶
sequence ['si:kwəns]	n. 序列,顺序
experimentally [ɪkspərɪ'mentəlɪ]	adv. 实验地,通过实验地
ratio ['reɪʃɪəʊ]	n. 比例

amount [əˈmaʊnt]   n. 量
unity [ˈjuːnɪtɪ]   n. 单位
ribose [ˈraɪbəʊs]   n. 核糖
deoxyribose [diːˌɒksɪˈraɪbəʊs]   n. 脱氧核糖
van der Waals   范德华
contact [ˈkɒntækt]   n. 接触
postulate [ˈpɒstjʊleɪt]   v. 假设
mechanism [ˈmekənɪzəm]   n. 机理
genetic [dʒɪˈnetɪk]   adj. 遗传的

Notes

1. This structure has novel features which are of considerable biological interest.

句子分析：本句中 novel 是形容词，意为新奇的，which 引导定语从句，先行词为 features。of interest 等于 interesting，用"介词 of+名词"结构相当于形容词。在本句中意译为"有意义"。

译文：该结构具有新奇的特征，而这些特征有相当重要的生物学意义。

2. In other words, if an adenine forms one member of a pair, on either chain, then on these assumptions the other member must be thymine; similarly for guanine and cytosine.

句子分析：本句中 either chain 指 DNA 两条链中的任一条链，assumption 是 assume 的名词形式。本句揭示了 DNA 中碱基对的配对方式，即 A 与 T 配对，C 与 G 配对。

译文：换言之，在（DNA 双链的）任一条链上，如果腺嘌呤是碱基对中的一个碱基，那么根据该等假设，碱基对中的另一个碱基一定是胸腺嘧啶；类似地，鸟嘌呤与胞嘧啶也是如此。

3. It has been found experimentally that the ratio of the amounts of adenine to thymine, and the ratio of guanine to cytosine, are always very close to unity for deoxyribose nucleic acid.

句子分析：本句为主语从句，用 it 作形式主语，真正的主语是 that 引导的从句。句中 unity 就是指 1，即数量比接近 1。

译文：通过实验发现，对于脱氧核糖核酸而言，腺嘌呤与胸腺嘧啶的量之比，以及鸟嘌呤与胞嘧啶的量之比，总是很接近 1。

4. It has not escaped our notice that the specific pairing we have postulated immediately suggests a possible copying mechanism for the genetic material.

句子分析：本句中 not escaped our notice 即 we noticed，"我们注意到"的意思，改换表达方式，使行文不呆板。事实上，Watson 和 Crick 的这篇论文，除了阐述 DNA 的双螺旋结构之外，更为重要的是，提出了 DNA 精确复制的机理，这也是这篇论文最有价值的地方。

译文：我们注意到我们所提出的这种特异的碱基配对方式立刻提示了一种可能的遗传物质复制机理。

Reading Comprehension

1. Try to illustrate the double helix structure of DNA.

2. How many types of bases are there in DNA? What are the ratio of amounts between them? How do the base pairs form?

3. What's the significance of the specific base pairing the authors postulated?

Reading Material

Gene Cloning in Bacteria 细菌基因克隆

There are essentially four stages for gene cloning.

1. Preparation of the Gene

Cloning bacterial genes is generally achieved from total chromosomal DNA preparations ("shotgun" cloning) by cleaving the DNA with a restriction endonuclease that generates fragments of approximately 4 kilobase pairs (kb) each with a "sticky" complementary single stranded end. Eukaryotic genes contain introns that are not possessed in bacteria so DNA for cloning is generally obtained as a reverse transcriptase generated copy(c) DNA of the relevant mRNA. DNA may also be synthetically if the nucleotide or amino acid sequence is known.

2. Insertion into Vector

The vector is the replicon that will enable the gene to be maintained in the host cell and includes plasmids and phages for bacterial hosts. Plasmid vectors should have single sites for common restriction endonuclease and antibiotic resistance determinants that allow selection of transformants. The vector is cut with the same enzyme as that used to generate the chromosomal DNA fragments, and fragments and linearized vector are incubated with DNA ligase which covalently joins the DNA molecules. A heterogenous population of molecules results, including dimmers, trimers and multimers of fragment and recircularized plasmids. Some plasmids will contain an inserted fragment thus producing a hybrid, recombinant plasmid.

3. Transformation of Host Cells

The ligated plasmid mixture is introduced into bacterial cells specially treated so that they take up DNA in a process termed transformation. In most case E. coli is the preferred host with the advantages that calcium choride treated cells are highly transformable, the molecular biology of this bacterium is well understood and a variety of plasmid and phage vectors are available. E.coli transcribes and translates most Gram-positive and Gram-negative genes with the exception of some actinomycetes genes.

4. Detection of Cloned Gene

The insertion site within the vector is located within a tetracycline resistance gene. Transformants that are ampicillin resistant but tetracycline sensitive therefore represent cells containing a hybrid plasmid (insertion inactivation). Some common methods for identification of the desired clone include: (a) expression of the gene and direct detection of the product (for example an enzyme) or complementation of a mutation in the host; (b) immunological methods for screening of the desired product using specific antisera; and (c) colony hybridization to a radioactively labeled probe DNA of the desired gene if the gene is unlikely to be expressed. Once detected, the identity and structure of the gene is confirmed by mapping and DNA sequence analysis.

Words and Expressions

chromosomal [ˈkrəʊməsəʊməl]	adj. 染色体的
shotgun cloning	鸟枪法克隆
restriction endonuclease	限制性核酸内切酶
complementary [kɒmpləˈmentərɪ]	adj. 互补的
fragment [ˈfrægmənt]	n. 片段
Eukaryotic [juˈkærɪəʊtɪk]	adj. 真核细胞的
intron [ˈɪntrɒn]	n. 内含子
reverse transcriptase	逆转录酶
nucleotide [ˈnjuːklɪətaɪd]	n. 核苷酸
vector [ˈvektə]	n. 载体
replicon [ˈreplɪˌkɒn]	n. 复制子
host cell	宿主细胞
plasmid [ˈplæzmɪd]	n. 质粒
phage [feɪdʒ]	n. 噬菌体
transformant [trænsˈfɔːmənt]	n. 转化子
DNA ligase	DNA 连接酶
heterogenous [ˌhetəˈrɒdʒənəs]	adj. 异源的
recombinant [riːˈkɒmbənənt]	n. 重组
transcribe [trænsˈkraɪb]	vt. 转录
Gram-positive	adj. 革兰氏阳性的
Gram-negative	adj. 革兰氏阴性的
actinomycete [ˌæktɪnoʊˈmaɪˌsiːt]	n. 放线菌
tetracycline [tetrəˈsaɪklɪn, tetrəˈsaɪklaɪn]	n. 四环素
ampicillin [ˌæmpɪˈsɪlɪn]	n. 氨苄青霉素
resistant [rɪˈzɪstənt]	adj. 抗性的
sensitive [ˈsensɪtɪv]	adj. 敏感的
insertion inactivation	插入失活
mutation [mjuːˈteɪʃən]	n. 突变
immunological [ˌɪmjuːnɒˈlədʒɪkəl]	adj. 免疫学的
antisera	n. 抗血清（antiserum 的名词复数）
radioactively [ˌreɪdɪəʊˈæktɪvlɪ]	adv. 放射活性地
probe [prəʊb]	n. 探针
express [ɪksˈpres]	v. 基因表达
mapping [ˈmæpɪŋ]	n. 基因作图
DNA sequence analysis	DNA 序列分析

Lesson Three Biochemical Reaction
生化反应

Consider the simple reaction of nitrogen to make ammonia:

$$N_2 + 3H_2 \rightleftharpoons 2NH_3$$

About half of the world's production of ammonia is carried out industrially and half biologically. At first glance, the two processes look quite different. The industrial reaction takes place at 500℃ and uses gaseous hydrogen and a metal catalyst under high pressure. The biochemical reaction takes place in the soil, uses bacterial or plant reactors, and occurs at moderate temperature and normal atmospheric pressure of nitrogen. These differences are so substantial that, historically, they were interpreted by supposing that biological systems are infused with a vital spirit that makes life possible. However, the biochemical reaction can be done with a purified enzyme. Like almost all biochemical reactions, the biological synthesis of ammonia requires a specific biochemical catalyst—an enzyme—to succeed. Enzymes are usually proteins and usually act as true catalysts; they carry out their reactions many times. The biological reduction of nitrogen is more similar to than different from its industrial counterpart: the energy change from synthesis of a mole of ammonia is identical in both cases, the substrates are the same, and the detailed chemical reaction is similar whether the catalyst is a metal or the active site of an enzyme.

Although there are many possible biochemical reactions, they fall into only a few types to consider: 1) oxidation and reduction, for example, the interconversion of an alcohol and an aldehyde; 2) movement of functional groups within or between molecules, for example, the transfer of phosphate groups from one oxygen to another; 3) addition and removal of water, for example, hydrolysis of an amide linkage to an amine and a carboxyl group; 4) bond-breaking reactions, for example, carbon-carbon bond breakage.

The complexity of life results, not from many different types of reactions, but rather from these simple reactions occurring in many different situations. Thus, for example, water can be added to a double bond as a step in the breakdown of many different compounds, including sugars, lipids, and peptides.

Words and Expressions

ammonia ['æməʊnjə]	n. 氨
catalyst ['kætəlɪst]	n. 催化剂
bacterial [bæk'tɪərɪrl]	n. 细菌
atmospheric pressure	气压
vital spirit	生命的精气
enzyme ['enzaɪm]	n. 酶
protein ['prəʊtiːn]	n. 蛋白质

biological synthesis	生物合成
reduction [rɪ'dʌkʃən]	n. 还原
mole [məʊl]	n. 摩尔
substrate ['sʌbstreɪt]	n. 底物
active site of an enzyme	酶的活性部位
oxidation and reduction	氧化和还原
interconversion [ˌɪntɜ:kən'vɜ:ʃn]	n. 相互转化
alcohol ['ælkəhɒl]	n. 醇
aldehyde ['ældɪhaɪd]	n. 醛
functional group	官能团
phosphate groups	磷酸基团
addition [ə'dɪʃən]	n. 加合
removal [rɪ'mu:vəl]	n. 消去
hydrolysis [haɪ'drɒlɪsɪs]	n. 水解
amide ['æmaɪd]	n. 酰胺
amine ['æmi:n]	n. 胺
carboxyl group	羧基
bond-breaking reactions	键断裂反应
carbon ['kɑ:bən]	n. 碳
double bond	双键
lipid ['lɪpɪd, 'laɪpɪd]	n. 脂
peptide ['peptaɪd]	n. 多肽

Notes

1. The biochemical reaction takes place in the soil, uses bacterial or plant reactors, and occurs at moderate temperature and normal atmospheric pressure of nitrogen.

句子分析：uses bacterial or plant reactors 指该生物化学反应发生的场所为细菌内或植物内，例如豆科植物的根瘤菌即具有生物固氮功能。normal atmospheric pressure of nitrogen 为介词 at 的宾语，意为通常气压的氮气，翻译时应注意句子的结构分析。

译文：该生物化学反应发生在土壤中，使用细菌或植物作为反应器，且反应在常温和常压下进行。

2. These differences are so substantial that, historically, they were interpreted by supposing that biological systems are infused with a vital spirit that makes life possible.

句子分析：本句为 so...that... 引导的结果状语从句。vital spirit 意为"生命的精气"或"生命力"，是一种生物学史上的错误观点即认为有机体含有特殊的生命力，是对有机体和无生命物体之间的差异的错误解释。

译文：这些差异是如此巨大，因此历史上曾经有人认为生物系统是因为具有"生命的精气"才使得生命现象成为可能。

3. The biological reduction of nitrogen is more similar to than different from its industrial counterpart: the energy change from synthesis of a mole of ammonia is identical in both cases, the

substrates are the same, and the detailed chemical reaction is similar whether the catalyst is a metal or the active site of an enzyme.

句子分析：More A than B，词组，意为"与其说是 B，毋宁说是 A"，或"是 A 而非 B"。Substrate 本意指酶的底物，在句中也指化学反应中的反应物。whether the catalyst is a metal or the active site of an enzyme 意为不管起催化作用的是金属催化剂（化学反应）还是酶的活性部位（生物化学反应）。

译文：氮的生物还原类似于而非区别于工业上氮的还原：在两种情形下生成一摩尔氨的能量变化是相同的，反应物是相同的，而且不管催化剂是金属还是酶的活性部位，该具体的化学反应都是类似的。

4. The complexity of life results, not from many different types of reactions, but rather from these simple reactions occurring in many different situations.

句子分析：根据原文，要进行适当增译，才能把原文的意思表达清楚。本句中 not…but rather…，即"不是……而是"的意思，A results from B，意为"A 由 B 产生，B 是 A 的原因"。

译文：生命的复杂性不是（有机体具有）很多不同类型的反应的结果，而是（有机体内）这些简单的反应在很多不同的情形下发生的结果。

Reading Material

Enzymes 酶

Enzymes

A cup of sugar left undisturbed for twenty years will change very little. But when you put some of the sugar in your mouth, it undergoes chemical change very rapidly. Enzymes secreted by some of your cells account for the difference in the rate of change, Enzymes are proteins with enormous catalytic power, which means they greatly enhance the rate at which specific reactions approach equilibrium.

Enzymes do not make anything happen that would not eventually happen on its own. They merely make it happen more rapidly (at least a million times faster, usually), and they make it happen again and again; enzyme molecules are not permanently altered or consumed in a reaction.

Also, an enzyme is quite selective about which reaction it will enhance and which reactant it will deal with, these being called substrates. For example, thrombin (an enzyme involved in blood clotting) catalyzes the breaking of a peptide bond only when the bond is between two particular amino acids: arginine and glycine.

Why is the preference of enzymes for specific substrates so important? If we think of metabolic pathways as chemical roads through a cell, then enzymes are like on-ramps at crossroads and traffic lights along certain routes. They allow only specific substrates to enter a given sequence of reactions, and they keep substrates moving through the sequence.

Controls over the enzymes of different pathways allow cells to direct the flow of nutrients, building materials, waste products, hormones, and so on in suitable ways. When you eat too much sugar, enzymes in your liver cells act on the excess, converting it first to glucose and then to glycogen or fat. When your body uses up glucose and needs more, enzymes break down glycogen to release its glucose subunits. In this case, a hormone called glucagons acts as a control over enzyme activity. It stimulates the key enzyme in the pathway by which glycogen is degraded, and it inhibits the enzyme that catalyzes glycogen formation.

Words and Expressions

undisturbed [ˈʌndɪsˈtɜːbd]	adj. 不受打扰的
undergo [ˌʌndəˈgəʊ]	v. 经历
secrete [sɪˈkriːt]	v. 分泌
enhance [ɪnˈhɑːns]	v. 促进，提高
approach [əˈprəʊtʃ]	v. 接近
equilibrium [ˌiːkwɪˈlɪbrɪəm]	n. 平衡
eventually [ɪˈventjʊəlɪ]	adv. 最终
permanently [ˈpɜːməntlɪ]	adv. 永久地
alter [ˈɔːltə]	v. 改变
consume [kənˈsjuːm]	v. 消耗
substrate [ˈsʌbstreɪt]	n. 底物
thrombin [ˈθrɒmbɪn]	n. 凝血酶
clot [klɒt]	v. 凝结
catalyze [ˈkætəlaɪz]	v. 催化
peptide bond	肽键
amino acid	氨基酸
arginine [ˈɑːdʒəˌniːn]	n. 精氨酸
glycine [ˈglaɪsiːn]	n. 甘氨酸
preference [ˈprefərəns]	n. 偏爱
metabolic pathway	代谢途径
nutrient [ˈnjuːtrɪənt]	n. 营养成分
hormone [ˈhɔːməʊn]	n. 激素
liver [ˈlɪvə]	n. 肝脏
convert [kənˈvɜːt, ˈkɒnvɜːt]	v. 转化，转变
glucose [ˈgluːkəʊs]	n. 葡萄糖
glycogen [ˈglaɪkəʊdʒen]	n. 糖原
break down	分解
subunit [sʌbˈjuːnɪt]	n. 亚基
glucagon [ˈgluːkəˌgɒn]	n. 胰高血糖素
stimulate [ˈstɪmjʊleɪt]	v. 刺激
inhibit [ɪnˈhɪbɪt]	v. 抑制
formation [fɔːˈmeɪʃən]	n. 形成，生成

History, Inheritance and Development
The Creation of Synthetic Crystalline Bovine Insulin

More than fifty years ago, a great achievement in life science occurred in China—the complete synthesis of crystalline bovine insulin—which gave Chinese scientists a sense of great elation and pride. Insulin is a hormone secreted by β cells in pancreas. Before the clinical application of insulin, diabetes was a feared disease that commonly led to death. Insulin has been studied since 1868 when Paul Langerhans, a medical student in Berlin found clusters of cells in the pancreas (Langerhans, 1868). These were later called "Islets of Langerhans". Some of these cells were eventually shown to produce insulin. The term "insulin" origins from "Insel", the German word for "islet" or "small island". Frederick Grant Banting, a young Canadian physician first extracted insulin from the pancreas of a dog whose pancreatic duct had been surgically ligated at University of Toronto in 1921, with the experimental facilities provided by Prof. John James Rickard Macleod and the assistance of one of Macleod's students, Charles H. Best. Biochemist James Bertram Collip helped purify the extract. For this work Banting and Macleod shared the 1923 Nobel Prize in Physiology or Medicine.

British molecular biologist Frederick Sanger determined the primary structure of insulin through 10 years of research: it comprises of two chains, chain A and chain B; chain A contains 21 amino acid residues while chain B consists of 30 residues; the two chains are linked by two disulfide bonds and there is an intra-chain disulfide bond in chain A. This is the first protein structure determined in human history, which Sanger was awarded the 1958 Nobel Prize in Chemistry for. Driven by the "great leap forward" campaign in 1958, Shanghai Institute of Biochemistry, Chinese Academy of Sciences and Perking University proposed that China should artificially synthesize insulin and obtained the support of the Chinese government. The project started in 1959, however, at that time, there was a lack of adequate equipment, the raw materials of amino acids and other necessary reagents. Consequently, synthesis of such a large compound represented a formidable task. The strategy adopted was to involve as many capable scientists as possible with eventually several hundreds of participants from eight different institutes participating in the project. People worked day and night preparing amino acids and other reagents, purifying solvents and synthesizing small peptides.

The synthesis of a complete and active insulin in only six years was a fantastic achievement given that it was predicted to take much longer. Why was such a brilliant and awesome achievement first accomplished in China, a developing country where the basis of scientific research was relatively weak, rather than in developed countries such as America and Germany? Besides the timely decision-making and strategic planning of the scientific administrative department in the government, the most important determinants might be the laudable mentality of Chinese scientists at that time, which can be summed up into "insulin spirit", which includes four aspects: (1) selfless dedication. All people involved in the project devoted all themselves into the demands of the project without considering their own interest; (2) honesty. Every intermediate in more than 200 steps of the synthetic procedure had to be rigorously identified, so even a slight problem on these identification procedures might lead to total failure of the whole project; (3) close cooperation. The three teams organized from three different institutes had clear assignment of their responsibilities and worked synergistically for the common aim of the project, so that high efficiency was achieved; (4) the spirit of welcoming challenges. To synthesize a protein consisting of 51 amino acids was a formidable task, so it was the spirit of welcoming a challenge that helped the Chinese scientists gain the respect of the world. Today, the "insulin spirit" is still of great value in constructing our research systems and managing research projects. Most importantly, it has become the source of confidence and strength of every Chinese researcher.

Practice and Training

Responsible Care

Started in Canada in 1985, Responsible Care is a global, voluntary initiative developed autonomously by the chemical industry for the chemical industry. It runs in 67 countries whose combined chemical industries account for nearly 90% of global chemical production. 96 of the 100 largest chemical producers in the world have adopted Responsible Care.

It stands for the chemical industry's desire to improve health, safety, and environmental performance.

The signatory chemical companies agree to commit themselves to improve their performances in the fields of environmental protection, occupational safety and health protection, plant safety, product stewardship and logistics, as well as to continuously improve dialog with their neighbors and the public, independent from legal requirements.

As part of Responsible Care initiative, the International Council of Chemical Associations introduced the Global Product Strategy in 2006.

Critical analyses of Responsible Care have been done by Andrew King and Michael Lenox Michael Givel, and Moffet, Bregha and Middelkoop.

History

Responsible Care was launched by the Chemistry Industry Association of Canada (formerly the Canadian Chemical Producers' Association - CCPA) in 1985. The term was coined by CIAC president Jean Bélanger. The scheme evolved, and, in 2006, The Responsible Care Global Charter was launched at the UN-led International Conference on Chemicals Management in Dubai.

Controversy

Poisoning of river Elbe by the company Draslovka Kolin a.s., a Responsible Care certified company or hiding of 20 metric tons leakage of naphthalene by the company Deza, also a Responsible Care company, questions whether the Responsible Care brings any real improvement or is just a marketing tool.

Unit Nine

Environmental Pollution and Control

环境污染及治理

Sewage treatment

Lesson One   Water Pollution and Pollutants
水污染及污染物质

The relationship between polluted water and disease was firmly established with the cholera epidemic of 1854 in London, England. Protection of public health, the original purpose of pollution control, continues to be the primary objective in many areas. However, preservation of water resources, protection of fishing areas, and maintenance of recreational waters are additional concerns today. Water pollution problems intensified following World War II when dramatic increases in urban density and industrialization occurred. Concern over water pollution reached a peak in the mid-seventies.

Water pollution is imprecise term that reveals nothing about either the type of polluting material or its source. The way we deal with the waste problem depends upon whether the contaminants are oxygen demanding, algae promoting, infectious, toxic, or simply unsightly. Pollution of our water resource can occur directly from sewer outfalls or industrial discharges (point sources) or indirectly from air pollution or agricultural or urban runoff (nonpoint sources).

Chemically pure water is a collection of H_2O molecules—nothing else. Such a substance is not found in nature—not in wild streams or lakes, not in clouds or rain, not in falling snow, nor in the polar ice caps. Very pure water can be prepared in the laboratory but only with considerable difficulty. Water accepts and holds foreign matter.

Wastewater is mostly water by weight. Other materials make up only a small portion of wastewater, but can be present in large enough quantities to endanger public health and the environment. Because practically anything that can be flushed down a toilet, drain, or sewer can be found in wastewater, even household sewage contains many potential pollutants. The wastewater components that should be of most concern to homeowners and communities are those that have the potential to cause disease or detrimental environmental effects.

Organisms

Many different types of organisms live in wastewater and some are essential contributors to treatment. A variety of bacteria, protozoa, and worms work to break down certain carbon-based (organic) pollutants in wastewater by consuming them. Through this process, organisms turn wastes into carbon dioxide, water, or new cell growth. Bacteria and other microorganisms are particularly plentiful in wastewater and accomplish most of the treatment. Most wastewater treatment systems are particularly plentiful in wastewater and accomplish most of the treatment. Most wastewater treatment systems are designed to rely in large part on biological processes.

Pathogens

Many disease-causing viruses, parasites, and bacteria also present in wastewater and enter from almost anywhere in the community. These pathogens often originate from people and animals that are infected with or are carriers of a disease. Graywater from typical homes contain enough pathogens to pose a risk to public health. Other likely sources in communities include hospitals, schools, farms, and food processing plants.

Some illnesses from wastewater-related sources are relatively common. Gastroenteritis can result from a variety of pathogens in wastewater, and cases of illnesses caused by the parasitic protozoa Guardian lambda and Cryptosporidium are not unusual in the US. Other important wastewater-related diseases include hepatitis A, typhoid, polio, cholera, and dysentery. Outbreaks of these diseases can occur as a result of drinking water from wells polluted by wastewater, eating contaminated fish, or recreational activities in polluted waters. Animals and insects that come in contact with wastewater can spread some illnesses.

Even municipal drinking water sources are not completely immune to health risks from wastewater pathogens. Drinking water treatment efforts can become overwhelmed when water resources are heavily polluted by wastewater. For this reason, wastewater treatment is as important to public health as drinking water treatment.

Organic Matter

Organic materials are found everywhere in the environment. They are composed of the carbon-based chemicals that are the building blocks of most living things. Organic materials in wastewater originate from plants, animals, or synthetic organic compounds, and enter wastewater in human wastes, paper products, detergents, cosmetics, foods, and from agricultural, commercial, and industrial sources.

Organic compounds normally are some combination of carbon hydrogen, oxygen, nitrogen, and other elements. Many organics are proteins, carbohydrates or fats and are biodegradable, which means they can be consumed and broken down by organisms. However, even biodegradable materials can cause pollution. In fact, too much organic matter in wastewater can be devastating to receiving waters.

Large amounts of biodegradable materials are dangerous to lakes, streams, and oceans, because organisms use dissolved oxygen in the water to break down the wastes. This can reduce or deplete the supply of oxygen in the water needed by aquatic life, resulting in fish kills, odors, and overall degradation of water quality. The amount of oxygen organisms need to break down wastes in wastewater is referred to as the biochemical oxygen demand (BOD) and is one of the measurements used to assess overall wastewater strength. Some organic compounds are more stable than others and cannot be quickly broken down by organisms, posing an additional challenge for treatment. This is true of many synthetic organic compounds developed for agriculture and industry.

In addition, certain synthetic organics are highly toxic. Pesticides and herbicides are toxic to humans, fish, and aquatic plants and often are disposed of improperly in drains or carried in storm water. In receiving waters, they kill or contaminate fish, making them unfit to eat. They also can damage processes in treatment plants. Benzene and toluene are two toxic organic compounds found in some solvents, pesticides, and other products. New synthetic organic compounds are being developed all the time, which can complicate treatment efforts.

Oil and Grease

Fatty organic materials from animals, vegetables, and petroleum also are not quickly broken down by bacteria and can cause pollution in receiving environments. When large amounts of oils and greases are discharged to receiving waters from community systems, they increase BOD and they may float to the surface and harden, causing aesthetically unpleasing conditions. They also can trap trash, plants, and other materials, causing foul odors, attracting flies and mosquitoes and other disease vectors. In some cases, too much oil and grease causes septic conditions in ponds and lakes by preventing oxygen from the atmosphere from reaching the water.

Onsite systems also can be harmed by harmed by too much oil and grease, which can clog onsite system drain field pipes and soils, adding to the risk of system failure. Excessive grease also adds to the septic tank scum layer, causing more frequent tank pumping to be required. Both possibilities can result in significant costs to homeowners. Petroleum-based water oils used for motors and industry are considered hazardous waste and should be collected and disposed of separately from wastewater.

Inorganics

Inorganic minerals, metals, and compounds, such as sodium, potassium, calcium, magnesium, cadmium, copper, lead, nickel, and zinc are common in wastewater from both residential and nonresidential sources. They can originate from a variety of sources in the community including industrial and commercial sources, storm water, and inflow and infiltration from cracked pipes and leaky manhole covers. Most inorganic substances are relatively stable, and cannot be broken down easily by organisms in wastewater.

Large amounts of many inorganic substances can contaminate soil and water. Some are toxic to animals and humans and may accumulate in the environment. For this reason, extra treatment steps are often required to remove inorganic materials from industrial wastewater sources. For example, heavy metals that are discharged with many types of industrial wastewater are difficult to remove by conventional treatment methods. Although acute poisonings from heavy metals in drinking water are rare in the US, potential long-term health effects of ingesting small amounts of some inorganic substances over an extended period of time are possible.

Nutrients

Wastewater often contains large amounts of the nutrients nitrogen and phosphorus in the form of nitrate and phosphate, which promote plant growth. Organisms only require small amounts of nutrients in biological treatment, so there normally is excess available in treated wastewater. In severe cases, excessive nutrients in receiving waters cause algae and other plants to grow quickly depleting oxygen in the water. Deprived of oxygen, fish and other aquatic lives die, and emitting foul odors.

Nutrients from wastewater have also linked to ocean "red tides" that poison fish and cause of a serious illness in humans. Nitrogen in drinking water may contribute to miscarriages and is the cause of a serious illness in infants called methemoglobinemia or "blue baby syndrome".

Solids

Solid materials in wastewater can consist of organic or inorganic materials and organisms. The solids must be significantly reduced by treatment or they can increase BOD when discharged to receiving waters and provide places for microorganisms to escape disinfection. They also can clog soil absorption fields in onsite systems. Solids can be divided into three states as follows:

1. Settleable solids-certain substances, such as sand, grit, and heavier organic and inorganic materials settle out from the rest of the wastewater stream during the preliminary stages of treatment. On the bottom of settling tanks and ponds, organic material makes up a biologically active layer of sludge that aids in treatment.

2. Suspended solids-materials that resist settling may remain suspended in wastewater must be treated, or they will clog soil absorption systems or reduce the effectiveness of disinfection's systems.

3. Dissolved solids-small particles of certain wastewater materials can dissolve like salt in water. Some dissolved materials are consumed by microorganisms in wastewater, but others, such as heavy metals, are difficult to remove by conventional treatment. Excessive amounts of dissolved solids in wastewater can have adverse effects on the environment.

Gases

Certain gases in wastewater can cause odors, affect treatment, or are potentially dangerous. Methane gas, for example, is a byproduct of anaerobic biological treatment and is highly combustible. Special precautions need to be taken near septic tanks, manholes, treatment plants, and other areas where wastewater gases can collect.

The gases hydrogen sulfide and ammonia can be toxic and pose asphyxiation hazards. Ammonia as a dissolved gas in wastewater also is dangerous to fish. Both gases emit odors, which can be a serious nuisance. Unless effectively contained or minimized by design and location, wastewater odors can affect the mental well being and quality of life of residents. In some cases, odors can even lower property values and affect the local economy.

Words and Expressions

cholera ['kɒlərə]	n. 霍乱
epidemic [ˌepɪ'demɪk]	adj. 流行（的），传染（的）
contaminant [kən'tæmənənt]	n. 污染物，致污物
algae ['ældʒiː]	n. 海藻，藻类（alga 的复数）
unsightly [ʌn'saɪtlɪ]	adj. 难看的，不雅观的
outfall ['aʊtfɔːl]	n. 出口，排口，河口
runoff ['rʌnˌɔf]	n. 径流

detrimental [ˌdetrɪˈmentl]	adj.	有害的，不利的
protozoa [ˌproutəˈzouə]	n.	原生动物（protozoan 的复数）
worm [wɜːm]	n.	蛆，寄生虫
microorganism [ˌmaɪkroˈɔːrəˌnɪzəm]	n.	微生物，微小动植物
viruses [ˈvaɪərəsɪz]	n.	病毒（virus 的复数）
parasite [ˈpærəsaɪt]	n.	食客，寄生虫
pathogen [ˈpæθədʒ(ə)n]	n.	病原体，病菌
gastroenteritis [ˌgæstrəʊˌentəˈraɪtɪs]	n.	肠胃炎
cryptosporidium [ˌkrɪptoʊspəˈrɪdɪrm]	n.	隐孢子虫
hepatitis [ˌhepəˈtaɪtɪs]	n.	肠炎
typhoid [ˈtaɪfɔɪd]	adj. 伤寒的； n. 伤寒	
polio [ˈpəʊlɪəʊ]	n.	小儿麻痹症（即 poliomyelitis）
dysentery [ˈdɪsəntrɪ]	n.	痢疾
detergent [dɪˈtɜːdʒənt]	n.	清洁剂，去垢剂
pesticide [ˈpestɪsaɪd]	n.	杀虫剂
herbicide [ˈhɜːbɪsaɪd]	n.	除草剂
immune [ɪˈmjuːn]	adj.	免疫的,不受影响的
cosmetic [kɒzˈmetɪk]	adj. 美容的； n. 化妆品，装饰品	
carbohydrate [ˈkɑːbəʊˈhaɪdreɪt]	n.	糖类，碳水化合物
aquatic life		水生生物
organic compound		有机化合物
benzene [ˈbenziːn, benˈziːn]	n.	苯
toluene [ˈtɒljuiːn]	n.	甲苯
aesthetically [iːsˈθetɪkəlɪ]	adv.	审美地,美学观点上地
scum [skʌm]	n.	泡沫，浮渣
ammonia [əˈməʊnjə]	n.	氨，氨水
anaerobic [ˌænerəˈrəʊbɪk]	adj.	厌氧的
sodium [ˈsəʊdjəm, ˈsəʊdɪəm]	n.	钠（Na）
potassium [pəˈtæsɪəm]	n.	钾（K）
calcium [ˈkælsɪəm]	n.	钙（Ca）
magnesium [mægˈniːzjəm]	n.	镁（Mg）
cadmium [ˈkædmɪəm]	n.	镉（Cd）
accumulate [əˈkjuːmiʊleɪt]	vi.	累积，积聚
methemoglobinemia [meθɪmoʊgloʊbɪˈniːmɪr]	n.	高铁血红蛋白症
methane [ˈmeθeɪn]	n.	甲烷，沼气
septic tank		化粪池
asphyxiation [æsˈfɪksɪəʃən]	n.	窒息

Notes

1. However, preservation of water resources, protection of fishing areas, and maintenance of

recreational waters are additional concerns today.

译文：然而，维护水资源、保护捕鱼区以及保养休闲用途水域，在当今仍是被关切的问题。

2. Outbreaks of these diseases can occur as a result of drinking water from wells polluted by wastewater, eating contaminated fish, or recreational activities in polluted waters.

译文：这些疾病的暴发原因是饮用了被废水污染了的井水、受污染的鱼类，或者是在污染了的水域进行了休闲活动。

3. The amount of oxygen organisms need to break down wastes in wastewater is referred to as the biochemical oxygen demand (BOD) and is one of the measurements used to assess overall wastewater strength.

译文：好氧生物降解废水中废物的含量指的是生物耗氧量，亦是总体评估废水浓度的测量方法之一。

4. When large amounts of oils and greases are discharged to receiving waters from community systems, they increase BOD and they may float to the surface and harden, causing aesthetically unpleasing conditions.

译文：当大量油脂从社区系统被排放到承受水体之后，它就会增加生物耗氧量，且会浮上水面并凝固，从而造成审美上的不愉悦。

Reading Comprehension

1. What is called chemically pure water?
2. What is water pollution's definition?
3. What ways do the oil and Grease influence the environment?

Reading Material

Special Considerations of the BODTrak™ Instrument
BODTrak™ 仪器的特殊说明

1. Range and Volume Selection

Because the BODTrak instrument can measure various waste substances, a variety of total oxygen demand test results can be obtained. The instrument provides four direct-reading ranges (0-35, 0-70, 0-350, and 0-700mg/L). Use Table 9-1 to select the volume and range corresponding to the expected sample BOD. For example, when using a sample with an expected BOD of 35mg/L or less, use a 420mL volume.

Table 9-1　Selection of sample volume

BOD range/（mg/L）	Required volume/mL
0-35	420
0-70	355
0-350	160
0-700	95

For best results, analyze samples immediately after collection. If this is not possible, preserve samples at low temperature (4℃) for no longer than 24 hours.

2. Sample Dilutions

If the sample's BOD is unknown, you can generally assume that effluent is normally in the 0-70mg/L range while influent is usually in the 0-700mg/L range.

If a sample does not contain sufficient nutrients for optimum bacteria growth, add the contents of one BOD Nutrient Buffer Pillow to each bottle. Do not add the BOD Nutrient Buffer Pillow if close simulation of original sample characteristics is required.

(1) When oxygen demand exceeds 700mg/L

When the O_2 demand of a sample exceeds 700mg/L, dilute the sample with high-quality dilution water. Make the dilution water with distilled water that does not contain organic matter or traces of toxic substances such as chlorine, copper, and mercury.

Demineralizers can release undetected organic matter that will create an objectionable oxygen demand. The most practical way to consistently produce water of low organic content is by distillation from alkaline permanganate. (For example, add 2g $KMnO_4$ and 4g NaOH for every liter of water.)

After distillation, place 3L of distilled water in a jug and bring the water temperature to 20℃. Add the contents of one BOD Nutrient Buffer Pillow for 3L to ensure sufficient nutrient concentration for diluted samples. Cap the jug and shake it vigorously for one minute to saturate the water with oxygen. Do not store the solution.

(2) Preparing several identical samples

Perform a single dilution for all samples when several identical samples are needed. After the dilution, multiply the reading by the dilution factor.

Example:

Prepare a 1:5 dilution by multiplying the original sample volume by 5 and adding dilution water until the new volume is obtained. If the sample volume is 200mL:

$$5 \times 200 = 1000 \text{ (mL)}$$

Dilute the 200mL sample to 1000 mL using the dilution water. Multiply the reading corresponding to the diluted sample by 5.

After sample dilution, refer to Table 10-1 to select volume and range.

3. Sample Seeding

(1) Determining BOD of seed

Certain types of BOD samples do not contain sufficient bacteria to oxidize the organic matter present in the sample. Many industrial discharges are of this type. Some sewage treatment plant effluents are chlorinated to the extent that they are essentially sterile, making it impossible to perform a direct BOD test. To test such samples, seed each bottle with water known to contain an abundant bacterial population.

The BOD of the seed must be known in order to determine the BOD of the sample. To determine the BOD of the seed, follow the same procedure used to determine the BOD of the sample. Run a BOD test on the pure seed and sample at the same time.

(2) Determining BOD of sample

After determining the BOD of the seed, apply the following formula to determine the sample's BOD.

$$\text{BOD sample} = \frac{(\text{BOD observed}) - (\text{Decimal fraction of seed used} \times \text{BOD seed})}{\text{Decimal fraction of sample used}}$$

Example:

A seeded sample is 10% seed and 90% sample (by volume).

The observed BOD is 60mg/L, and the pure seed BOD is 150mg/L.

$$\text{BOD sample} = \frac{(60 \text{ mg/L}) - (0.10 \times 150 \text{ mg/L})}{0.90} = 50 \text{ mg/L}$$

(3) Variations in initial bacterial populations

Microbic counts of domestic sewage fluctuate by the hour, so bacteria counts should be reported within a specified range. Domestic sewage typically contains between 10^4 and 10^6 cells per mL. Variations in initial bacterial populations will have a minimal effect on the BOD test if the bacteria count in the sample exceeds 10^3 cells per mL.

Low seed concentrations are more critical than those that are too high. They delay the start of oxidation and cause low BOD results. Use the trial and error method to determine the optimum concentration of seed for a specific waste material. Adding 10% of Polyseed Solution or raw domestic sewage seed to each BOD sample is usually sufficient for the manometric method.

Try various concentrations of seed and determine the respective BODs of the waste sample as well as of the seed itself. Choose the seed concentration yielding the highest corrected waste sample BOD. This seed percentage can range from 2% to 30%, depending on the waste material tested.

4. Sample Temperature

The American Public Health Association (APHA) recommends a solution temperature of 20℃ ±1℃ (68°F) for conducting the BOD test. Obtain this temperature by placing the BODTrak instrument in an appropriate incubator and adjusting the temperature until the solution reaches 20℃ ±1℃. An undercounter BOD Incubator and a combination BOD Incubator/Refrigerator are available (see PARTS & ACCESSORIES).

Samples should be cooled to incubation temperature. Seeding of samples with initially high sample temperatures may also be necessary because samples may have insufficient bacteria. Determine the BOD of the seed and the BOD of the sample at the same time.

5. Industrial Wastes

Industrial and chlorinated samples often contain toxic substances and require special considerations when running BOD tests. The presence of toxic substances in the sample will cause decreased BOD values. Either remove the toxic substances or eliminate their effects by diluting the sample.

(1) Chlorine

Low chlorine concentrations may be dissipated by maintaining the sample at room temperature for 1 to 2 hours before testing. Remove the chlorine from samples with high chlorine concentration by adding sodium thiosulfate as described below:

① Add 10mL of 0.02N Sulfuric Acid Standard Solution and 10mL of 100mg/L Potassium Iodide Solution to a 100mL portion of sample in a 250mL Erlenmeyer flask.

② Add three droppers of Starch Indicator Solution and swirl to mix.

③ Titrate from dark blue to colorless with 0.025N Sodium Thiosulfate Standard Solution.

④ Calculate the amount of Sodium Thiosulfate Standard Solution necessary to dechlorinate the remaining sample:

$$\text{mL of Sodium Thiosulfate} = \frac{(\text{mL used})(\text{mL sample to be dechlorinated})}{100}$$

⑤ Add the required amount of 0.025N Sodium Thiosulfate Standard Solution to the sample and mix thoroughly. Wait 10 to 20 minutes before running the BOD test.

(2) Other toxic materials

Determine the concentrations of other toxic materials such as phenols, heavy metals, and cyanides.

NOTE: To determine the concentrations of these and other materials, see the Hach "Water Analysis Handbook."

Dilute the sample with distilled water to eliminate the effect of these materials. The correct BOD is obtained when two successive dilutions result in the same sample BOD value.

(3) Seed acclimatization

Domestic sewage or Polyseed Inoculum can provide seed for most samples. Polyseed Inoculum is ideally suited for domestic and industrial wastewater because it provides a constant seed source and is free of nitrifying microorganisms.

Pour the contents of one polyseed capsule into dilution water to rehydrate (refer to the procedure packaged with the Polyseed). Aerate and stir for 1 hour. Use enough of this solution so that it makes up 10% to 30% of the overall sample volume. The exact percentage of seed must be determined for each sample type.

For more information, Standard Methods for the Examination of Water and Wastewater, 18th edition emphasizes the importance of selecting the proper seed for specific wastes.

If the waste sample to be tested contains toxic materials such as phenol, formaldehyde, or other microbic inhibitory agents, use acclimated seed. Acclimate the seed in any non-metal or stainless steel gallon container fitted with an aeration system. Proceed as follows:

① Aerate domestic sewage for about 24hours.Allow one hour settling time for heavier materials to settle.

② After the one hour settling, siphon and discard the top two-thirds of the volume.

③ Refill the container to the original volume with domestic sewage containing 10% of the waste material in question.

④ Repeat steps (1)-(3), increasing the addition of waste material by 10%. Stop the procedure when 100% waste material has been reached.

6. pH Effect

Low BOD test results occur when the pH of a test waste material exceeds the 6-8 range. The operator may maintain this pH to simulate original sample conditions or may adjust the pH to approach neutrality (buffered at pH 7). Neutralize samples containing caustic alkalinity or acidity by using 1.0N (or weaker) sulfuric acid or sodium hydroxide, respectively.

7. Supersaturation

Reduce supersaturated cold samples (containing more than 9mg/L dissolved oxygen at 20℃) to saturation. To do so, first bring the sample temperature to about 20℃. Then partly fill a sample bottle with sample and shake vigorously for two minutes, or aerate with filtered compressed air for two hours.

Lesson Two Air Pollution and Major Air Pollutants
大气污染及主要的大气污染物

Air pollution

As the fastest moving fluid medium in the environment, the atmosphere has always been one of the most convenient places to dispose of unwanted materials. Even since fire was first used by people, the atmosphere has been a sink for waste disposal.

1. Definition of Air Pollution

Air pollution can be defined as the presence in the outdoor atmosphere of one or more contaminants (pollution) in such quantities and of such duration as may be (or may tend to be) injurious to human, plant, or animal life, or to property (materials), or which may unreasonably interfere with the comfortable enjoyment of life or property, or the conduct of business. It should be stressed that the attention in this definition is on the outdoor, or ambient, air, as opposed to the indoor, or work environment, air. This definition mentions the quantity or concentration of the contaminant in the atmosphere, and its associated duration or period of occurrence. This is an important concept in those pollutants that are present at extremely low concentrations and for short time periods may be insignificant in terms of damage effect.

2. Sources of Air Pollutants

Air pollutant sources can be categorized from several perspectives, including the type of source, their frequency of occurrence and spatial distribution, and the types of emissions. Characterization by source type can be delineated as arising from natural sources or from man-made sources. "Natural sources" include plant pollens, windblown dust, volcanic eruptions, and lightning-generated forest fires. "Man-made sources" can include transportation vehicles, industrial processes, power plants, construction activities, and military training activities. The research date suggest that, with the exception of sulfur and nitrogen oxides, natural emissions of air pollutants exceed human-produced input. Nevertheless, it is the human component that is most abundant in urban areas and that leads to the most severe air pollution problems for human health.

Source characterization according to spatial distribution can be categorized as stationary sources and mobile sources. Stationary sources are those that have a relatively fixed location. These include point sources, fugitive sources, and area sources. Point sources are stationary sources that emit air pollutants from one or more controllable sites, such as smokestacks of power plants at industrial sites. Fugitive sources are types of stationary sources that generate air pollutants from open areas exposed to wind processes. Examples include dirt roads, construction sites, farmlands, storage piles, surface mines, and other exposed areas from which particulates may be removed by wind. Area sources are locations from which air pollutants are emitted from a well-defined area within which are several sources, for example, small urban communities or areas of intense industrialization within urban complexes or agricultural areas sprayed with herbicides and pesticides. Mobile sources are emitters of air pollutants that move from place to place while yielding emissions. These include automobiles, trucks, aircraft, ships, and trains.

3. Major Air Pollutants

(1) Sulfur dioxide

Sulfur Dioxide(SO_2) is a colorless and odorless gas normally present at earth's surface at low concentrations. One of the significant features of SO_2 is that once it is emitted into the atmosphere it may be converted through complete reactions to fine particulate sulfate (MSO_4). The major anthropogenic source of sulfur dioxide is the burning of fossil fuels, mostly coal in power plants. Another major source comprises a variety of industrial processes, ranging from petroleum refining to the production of paper cement, and aluminum.

Adverse effects associated with sulfur dioxide depend on the dose or concentration present and include corrosion of paint and metals and injury or death to animals and plants. Crops such as alfalfa, cotton, and barley are especially susceptible. Sulfur dioxide is capable of causing severe damage to human and other animal lungs, particularly in the sulfate form. It is also an important precursor to acid rain, as are nitrogen oxides.

(2) Nitrogen oxides

Nitrogen oxides (NO_x) are emitted largely in two forms: nitric oxides (NO) and nitrogen dioxide (NO_2) (the x in NO_x refers to the number of oxygen atoms present in the gas molecule). Although nitrogen oxides occur in many forms in the atmosphere, NO and NO_2 are subject to emission regulations only. The most important of these is NO_2, which is a visible yellow brown to reddish brown gas. A major concern with nitrogen dioxide is that it may be converted by complex reactions in the atmosphere to the ion (NO_3^-) within small water particles, impairing visibility. Additionally, nitrogen dioxide is one of the main pollutants that contribute to the development of smog (as is NO) and is a major contributor to acid rain. Nearly all NO_x is emitted from anthropogenic sources; the two major contributors are automobiles and power plants that burn fossil fuels.

The environmental effects of nitrogen oxides on humans are variable but include the irritation of eyes, nose, throat, and lungs and increased susceptibility to viral infections, including influenza (which can cause bronchitis and pneumonia). Nitrogen oxides suppress plant growth and damage leaf tissue. When the oxides are converted to their nitrate form in the atmosphere, they impair visibility. However, when nitrate is deposited on the soil, it can promote plant growth through nitrogen fertilization.

(3) Carbon monoxide

Carbon monoxide (CO) is a colorless, odorless gas that at very low concentrations is extremely toxic to humans and other animals. The high toxicity results from a striking physiological effect, namely, that carbon monoxide and hemoglobin in blood have a strong natural attraction for one another. Hemoglobin in our blood will take up carbon monoxide nearly 250 times more rapidly than it will oxygen. Therefore, if there is any carbon monoxide in the vicinity, a person will take it in very readily, with potentially dire effects. Many people have been accidentally asphyxiated by carbon monoxide produced from incomplete combustion of fuels in campers, tents, and houses. The effects depend on the dose or concentration of exposure and range from dizziness and headaches to death. Carbon monoxide is particularly hazardous to people with known heart disease, anemia, or respiratory disease. In addition, it may cause birth defects, including mental retardation and impairment of growth of the fetus. Finally, the effects of carbon, monoxide tend to be worse at higher altitudes, where oxygen levels are naturally lower.

Approximately 90% of the carbon monoxide in the atmosphere comes from natural sources, and the other 10% comes mainly from fires, automobiles, and other sources of incomplete burning of organic compounds. Concentrations of carbon monoxide can build up and cause serious health effects in a localized area.

(4) Photochemical oxidants

Photochemical Oxidants result from atmospheric interactions of nitrogen dioxide and sunlight. The most common photochemical oxidant is ozone (O_3), a colorless gas with a slightly sweet odor. In addition to ozone, a number of photochemical oxidants known as PANs (peroxyacyl nitrates) occur with photochemical smog.

Ozone is a form of oxygen in which three atoms of oxygen occur together rather than the normal two. Ozone is relatively unstable and releases its third oxygen atom readily, so that it oxidizes or burn thing more readily and at lower concentrations than does normal oxygen. Ozone is sometimes used to sterilize; for example, bubbling ozone gas through water is a method used to purify water. The ozone is toxic to and kills bacteria and other organisms in the water. When it is released into the air or produced in the air, ozone may injure living thing.

Ozone is very active chemically, and it has a short average lifetime in the air. Because of the effect of sunlight on normal oxygen, ozone forms a natural layer high in the atmosphere (stratosphere). This ozone layer protects us from harmful ultraviolet radiation from the sun. Ozone is considered a pollutant when present above the National Air Quality Standard threshold concentration in the lower atmosphere but is beneficial in the stratosphere.

The major sources of the chemicals that produce oxidants, and particularly ozone, are automobiles, fossil fuel burning, and industrial processes that produce nitrogen dioxide. Due to the natures of its formation, ozone is a difficult pollutant to regulate and is the pollutant whose health standard is most frequently exceeded in urban areas of the United States. The adverse environmental effects of ozone and other oxidants, as with other pollutants, depend in part on the dose or concentration of exposure and include damage to plants and animals as well as to materials such as rubber, paint, and textiles.

The effect of ozone on plants can be subtle. At very low concentrations, ozone can reduce growth rates while not producing any visible injury. At higher concentrations, ozone kills leaf tissue, eventually killing entire leaves and, if the pollutant levels remain high, killing whole plants. The death of white pine threes plants along highways in New England is believed due in part to ozone pollution.

Ozone's effect on animals, including people, involves various kinds of damage, especially to the eyes and the respiratory system.

(5) Hydrocarbons

Hydrocarbons are compounds composed of hydrogen and carbon. There are thousands of such compound, including natural gas or methane (CH_4), butane (C_4H_{10}), and propane (C_3H_8). Analysis of urban air has identified many different hydrocarbons some of which are much more reactive with sunlight (producing photochemical smog) than others. The potential adverse effects of hydrocarbons are numerous; many at a specific dose or concentration are toxic to plants and animals or may be covered to harmful compounds through complex chemical changes that occur in the atmosphere. Over 80% of (which are primary pollutants) that enter the atmosphere are emitted from natural sources. The most important anthropogenic source is the automobile. Hydrocarbons may also escape to the atmosphere when a car's tank is being filled with gasoline or gasoline is spilled and it evaporates. Vapor recovery systems on the hoses that feed the gasoline to the tank are now required in many urban areas and are helping to reduce the problem of hydrocarbons (vapors) escaping while tanks are being filled.

(6) Hydrogen sulfide

Hydrogen sulfide (H_2S) is a highly toxic and corrosive gas, easily identified by its rottenegg odor. Hydrogen sulfide is produced from natural sources, such as geysers, swamps, and bogs, as well as from human sources, such as petroleum processing-refining and metal smelting. The potential effects of hydrogen sulfide include functional damage to plants and health problems ranging from toxicity to death for humans and other animals.

(7) Hydrogen fluoride

Hydrogen fluoride (HF) is a gaseous pollutant that is released primarily by aluminum production, coal gasification, and the burning of coal in power plants. Hydrogen fluoride is extremely toxic, and even a small concentration (as low as 1×10^{-9}) may cause problems for plants and animals.

(8) Other hazardous gases

It is a rare month when the newspapers do not carry a story of a truck or train accident that releases toxic chemicals in a gaseous form into the atmosphere. In these incidents it is often necessary to evacuate people from the area until the leak is stopped or the gas dispersed to a nontoxic level. Chlorine gases are often the culprit, but a variety of other materials used in chemical and agricultural processes may be involved.

Some chemicals are so toxic that extreme care must be taken to ensure that they do not enter the environment. The danger of such chemicals was tragically demonstrated on December 3, 1984, when a toxic cloud chemical (stored in liquid form) at a pesticide plant leaked, vaporized, and formed a toxic cloud that settled over a $64km^2$ area of Bhopal, India. The gas leak lasted less than 1 hour, yet over 2000 people were killed and more than 15000 were injured by the gas, which causes severe irritation (burns on contact) to eyes, nose, throat, and lungs.

(9) Particulate matter

Particulate matter encompasses the small particles of solid or liquid substances that are released into the atmosphere by many activities. Modern farming adds considerable amounts of particulate matter to the atmosphere, as do desertification and volcanic eruptions. Nearly all industrial processes, as well as the burning of fossil fuels, release particulates into the atmosphere. Much particulate matter is easily visible as smoke, soot, or dust; other particulate matter is not easily visible. Included with the particulates are materials such as airborne asbestos particles and small particles of heavy metals, such as arsenic, copper, lead, and zinc, which are usually emitted form industrial facilities such as smelters.

Of particular importance with reference to particulates are the very fine particle pollutants less than 10μm in diameter （10 milionths of a meter）. For comparison, the diameter of human hair is about 60 μm to 150 μm. Among the most significant of the fine particulate pollutants are sulfates and nitrates. These are mostly secondary pollutants produced in the atmosphere through chemical reactions between normal atmospheric constituents and sulfur dioxide and nitrogen oxides. These reactions are particularly important in the formation of sulfuric and nitric acids in the atmosphere. When measured, particulate matter is often referred to as total suspended particulates (TSPs).

Particulates affect human health, ecosystems, and the biosphere. In the United alone, particulate air pollution contributes to the death of 6000 people annually. Recent studies of fine particulate pollution and health have estimated that 2% to 9% of total human mortality in cities is associated with particulate pollution; risk of mortality is approximately 15% to 25% higher in cities with the highest levels of fine particulate pollution compared to cities with the lowest levels. Particulates that enter the lungs may lodge there and have chronic effects on respiration. Certain materials, such as asbestos, are particularly dangerous in this way. Dust raised by road building and plowing and deposited on the surfaces of green plants may interfere with their absorption of carbon dioxide and oxygen and their release of water. Heavy dust may affect the breathing of animals. Particulates associated with large construction projects may kill organisms and damage large areas, changing species composition, altering food chains, and generally affecting ecosystems. In addition, modern industrial processes have greatly increased the total suspended particulates in earth's atmosphere. Particulates block sunlight and may cause changes in climate. Such changes have lasting effects on the biosphere.

(10) Asbestos

Asbestos is the term for several minerals that have the form of small elongated particles. In the past, asbestos was treated rather casually, and people working in asbestos plants were not protected from dust. Asbestos was used in building insulation, roofing material, and in brake pads for automobiles, trucks, and other vehicles. As a result, a considerable amount of asbestos fibers have been spread throughout industrialized countries, especially in urban environments of Europe and North America. In one case, the products containing asbestos were sold in burlap bags that were recycled by nurseries and other secondary businesses, thus further spreading the pollutant. Some types of asbestos particles are believed to be carcinogenic, or carry with them carcinogenic materials, and so must be carefully controlled.

(11) Lead

Lead is an important constituent of automobile batteries and other industrial products. When lead is added to gasoline, it helps protect the engine and promotes more even fuel consumption. The lead in gasoline is emitted into the environment in the exhaust. In this way, lead has been spread widely around the world and has high levels in soils and waters along roadways.

Once released, lead can be transported through the air as particulates to be taken up by plants through the soil or deposited directly on plant leaves. Thus it enters terrestrial food chains. When lead is carried by streams and rivers, deposited in quiet waters, or transported to ocean or lakes, it is taken up by aquatic organisms and thus enters aquatic food chains.

The concentration of lead measured in Greenland glaciers was essentially zero in A.D800 and reached measurable levels with the beginning of the industrial revolution in Europe. The lead content of the glacial ice increased steadily from 1750 until the mid-twentieth century (about 1950), when the rate of accumulation by the glaciers began to increase rapidly. This sudden upsurge reflects the rapid growth in the use of lead additives in gasoline. Lead reaches Greenland as airborne particulates and via seawater. The accumulation of lead in the Greenland ice illustrates that our use of heavy metals in this century has reached a point where the entire biosphere is affected. The reduction and eventual elimination of lead in gasoline is a good start. Lead has been removed from nearly all gasoline in the United States and Canada and is being phased out in much of Europe.

Words and Expressions

atmosphere ['ætməsfɪə]	n. 气氛，大气，空气
disposal [dɪs'pəuzəl]	n. 清理，处理
ambient ['æmbɪənt]	adj. 周围的，环绕的，外界的；n. 周围环境
windblown ['wɪnd,blon]	adj. 被风吹的，风飘型的
stationary source	固定污染源，固定来源
mobile source	流动污染源
smokestack ['smoukstæk]	n. 烟囱，烟窗
sulfur dioxide	二氧化硫
colorless ['kʌləlɪs]	adj. 无趣味的，苍白的，无色的
odorless ['əudəlɪs]	adj. 没有气味的
anthropogenic [,ænθrəpə'dʒnɪk]	adj. 人类起源的，人为的
corrosion [kə'rəuʒən]	n. 腐蚀，腐蚀产生的物质，衰败
sulfate ['sʌlfeɪt]	n. 硫酸盐；vt. 使成硫酸盐，用硫酸处理
influenza ['ɪnflu'enzə]	n. 流行性感冒（简写 flu）
carbon monoxide	一氧化碳
hemoglobin [,hi:məu'gləubɪn]	n. [生化]血红素，[生化]血红蛋白（haemoglobin）
vicinity [vɪ'sɪnɪtɪ]	n. 邻近，附近，近处
asphyxiate [æs'fɪksɪeɪt]	vt. 使……窒息；vi. 窒息
combustion [kəm'bʌstʃən]	n. 燃烧，氧化
dizziness ['dɪzɪnɪs]	n. 头晕，头昏眼花
photochemical oxidant	[化]光化学氧化剂

peroxyacyl nitrates	过氧乙酰硝酸酯
sterilize ['sterɪlaɪz]	vt. 消毒，杀菌
hydrocarbon ['haɪdrəʊ'kɑ:bən]	n. 碳氢化合物
butane ['bju:teɪn]	n. [化]丁烷
propane ['prəʊpeɪn]	n. [化]丙烷
hydrogen sulfide	硫化氢
corrosive [kə'rəʊsɪv]	adj. 腐蚀的，侵蚀性的；n. 腐蚀物
hydrogen fluoride	氟化氢
aluminum [ə'lju:mɪnəm]	n. 铝
chlorine ['klɔ:ri:n]	n. 氯
particulate matter	微粒物质，悬浮微粒
arsenic ['ɑ:sənɪk]	n. [化]砒霜，砷，三氧化二砷；adj. 含砷的
copper ['kɒpə]	n. 铜
zinc [zɪŋk]	n. 锌
ecosystem [i:kə'sɪstəm]	n. 生态系统
biosphere ['baɪəsfɪə]	n. 生物圈
lung [lʌŋ]	n. 肺，呼吸器
asbestos [æz'bestɒs]	n. [矿]石棉；adj. 石棉的
carcinogenic [kɑ:sɪnə'dʒənɪk]	adj. 致癌物的，致癌的
glacier ['glæsjə, 'gleɪʃə]	n. 冰河，冰川
airborne particulate	空中悬浮微粒

Notes

1. Air pollution can be defined as the presence in the outdoor atmosphere of one or more contaminants (pollutants) in such quantities and of such duration as may be (or may tend to be) injurious to human, plants, or animal life, or to property (materials), or which many unreasonably interfere with the comfortable enjoyment of life or property, or the conduct of business.

译文：空气污染可以定义为存在于室外大气中的一种或多种污染物，其数量和持续时间已达到（或将会）危害人类、动物、植物及财产的程度，或妨碍（人们）对生活财产的舒适享受或影响商业活动。

2. The environmental effects of nitrogen oxides on humans are variable but include the irritation of eyes, nose, throat, and lungs and increased susceptibility to viral infections, including influenza (which can cause bronchitis and pneumonia).

译文：氮氧化物对人类的影响不一，包括刺激眼睛、鼻子、喉咙和肺，以及增加人类对病毒的易感性，包括流感（可引起支气管炎和肺炎）。

3. The gas leak lasted less than 1 hour, yet over 2000 people were killed and more than 15000 were injured by the gas, which causes severe irritation (burns on contact) to eyes, nose, throat, and lungs.

译文：该毒气泄漏长达 1 小时之久，造成 2000 多人死亡，15000 多人受伤，并引发伤者眼、鼻、喉及肺部等多处严重刺痛。

Reading Comprehension

1. What is the definition of air pollution?
2. How is the classification of atmospheric pollutants?
3. What are the property and hazards of sulfur dioxide?
4. What are the effects of ozone on plants?

Reading Material

Acid Rain 酸雨

Acid rain is a broad term used to describe several ways that acids fall out of the atmosphere. A more precise term is acid deposition, which has two parts: wet and dry（Fig.9-1）.

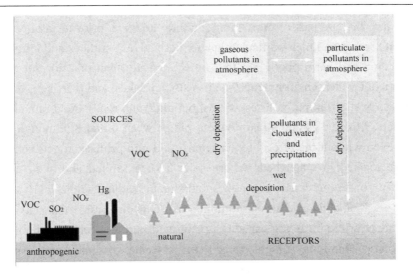

Acid rain

Fig. 9-1 The deposition pathways of acid rain

Wet deposition refers to acidic rain, fog, and snow. As this acidic water flows over and through the ground, it affects a variety of plants and animals. The strength of the effects depend on many factors, including how acidic the water is, the chemistry and buffering capacity of the soil involved, and the types of fish, tree, and other living things that rely on the water.

Dry deposition refers to acidic gases and particles. About half of the activity in the atmosphere falls back to earth through dry deposition. The wind blows these acidic particles and gases onto buildings, cars, homes and trees. Dry deposition gases and particles can also be washed form trees and other surfaces by rainstorm. When that happens, the runoff water adds those acidic to the acid rain, making the combination more acidic than the falling rain alone.

Prevailing wind blow the compounds that cause both wet and dry acidic deposition across state and national borders, and sometimes over hundreds of miles.

Scientists discovered, and have confirmed, that sulfur dioxide (SO_2) and nitrogen oxides (NO_x) are the primary causes of acid rain. In the US, About 2/3 of all SO_2 and 1/4 of all NO_x comes from electric power generation that relies on burning fossil fuels like coal.

How do we measure acid rain? Acid rain can be measured using a scale called pH. The lower a substances pH, the more acidic it is. See the pH page for more information.

Pure water has a pH of 7.0. Normal rain is slightly acidic because carbon dioxide dissolves into it, so it has a pH of about 5.5.

Recently, acid rain has attracted more and more attention of public, In the US, acid rain's pH and the chemicals that cause acid rain are monitored by two networks, both supported by EPA The National Atmospheric Deposition Program (NADP) measure wet deposition , The Clean Air Status and Trends Network (CASTNET) measure dry deposition .

Acid rain causes acidification of lakes streams and contributes to damage of trees at high elevations (for example, red spruce trees above 2000 feet) and many sensitive forest soils. In addition, acid rain accelerates the decay of building materials and paints, including irreplaceable, states, and sculpture that are part of our nation's cultural heritage. prior to falling to falling to the earth ,SO_2 and NO_x gases and their particulate matter derivatives , sulfates and nitrate, contribute to visibility degradation and harm public health. The ecological effect of acid rain are most clearly seen in the aquatic, or water, environments, such as stream, lakes, and marshes. Acid rain flows to streams, and marshes after falling on forest, fields, building, and roads. Acid rain also falls directly on aquatic habitats. Most lakes and stream have a pH between 6 and 8, although some lakes are naturally acidic even without the effects of acid rain. Acid rain primary affects sensitive bodies of water, which are located in watersheds whose soils have a limited ability to neutralize acidic compounds (called "buffering capacity"). Lakes and stream become acidic (pH value goes down) when the water itself and its and its surrounding soil con not buffer the acid rain enough to neutralize it. In areas where buffering capacity is low, acid rain also releases aluminum from soils into lakes and streams; aluminum is highly toxic to many species of aquatic organisms.

There are several ways to reduce acid deposition, more properly called acid deposition, ranging from societal changes to individual action. (1)Clean up smokestacks and exhaust pipes .Because almost all of the electricity that powers modern life comes from burning fossil fuels like coal, natural gas, and oil. While two pollutants that are released into the atmosphere, or emitted cause acid deposition, when there fuels are burned: sulfur dioxide (SO_2) and nitrogen oxides (NO_x). (2)Use alternative energy sources: There are other sources of electricity besides fossil fuels. They include nuclear power ,hydropower , wind energy ,geothermal energy , and solar energy .Of there, nuclear and hydropower are used most widely; wind, solar, and geothermal energy have not yet been harnessed on a large scale in this country. There are also alternative energies available to power automobiles, including natural gas powered vehicles. (3)Restore a damaged environment: Because there are so many changes, it takes many years for ecosystems to recover from acid deposition, even after emissions are reduced and the rain becomes normal again. However, there are some things that people do to bring back lakes and streams more quickly. For example limestone or lime can be added to acidic lakes to cancel out the acidity. This process, called liming, has been used extensively in Norway and Sweden. (4)Take action as individuals: It may seem like there is not much that one individual can do to stop acid deposition. However, like many environmental problems, acid deposition is caused by the cumulative actions of millions of individual people. Therefore, each individual can also reduce their contribution to the problem and become part of the solution.

Lesson Three   Sources and Types of Solid Wastes
固体废物的来源和类别

Knowledge of the sources and type of solid wastes, along with data on the composition and rates of generation, is basic to the design and operation of the functional elements associated with the management of solid wastes.

1. Sources of Solid Wastes

Sources of solid wastes are, in general, related to land use and zoning. Although any number of source classifications can be developed, the following categories have been found useful: (1) residential, (2)commercial, (3)municipal, (4)industrial, (5)open areas, (6)treatment plants, and (7)agricultural. Typical waste generation facilities, activities, or locations associated with each of these sources are presented in Table 9-2.The types of wastes generated, which are discussed next, are also identified.

Table 9-2   Typical solid waste generating facilities, activities, and locations associated with various source classifications

Source	Typical facilities, activities, or locations where wastes are generated	Types of solid wastes
Residential	Single-family and multifamily dwellings, low-, medium-, and high-rise apartments, etc.	Food wastes, rubbish, ashes, special wastes
Commercial Occasionally	Stores, restaurants, markets, office buildings, hotels, motels, motels, print shops, auto repair shops, medical facilities and institutions, etc.	Food wastes, rubbish, ashes, demolition and construction wastes, special wastes, hazardous wastes
Municipal*	As above*	As above*
Industrial demolition	Construction, fabrication, light and heavy manufacturing, refineries, chemical plants, lumbering, mining power plants, demolition, etc.	Food wastes, rubbish, ashes, and construction wastes, special wastes, hazardous wastes
Open areas	Streets, alleys, parks, vacant lots, playgrounds, beaches, highways, recreational areas, etc.	Special wastes, rubbish
Treatment plant sites principally	Water, waste water, and industrial treatment processes, etc.	Treatment plant wastes, composed of residual sludges
Agricultural	Field and row crops, orchards, vineyards, darries, foodlots, farms, etc.	Spoiled food wastes, agricultural wastes, rubbish, hazardous wastes

* The term municipal normally is assumed to include both the residential solid wastes generated in the community.

2. Type of Solid Wastes

The term solid waste is all-inclusive and encompasses all sources, type of classifications, composition, and properties. Wastes that are discharged may be of significant value in another setting, but they are of little or no value to the possessor who wants to dispose of them. To avoid confusion, the term refuse, often used interchangeably with the term solid wastes, is not used in this text.

As a basis for subsequent discussions, it will be helpful to define the various types of solid wastes that are generated (see Table 9-2). It is important to be aware that the definitions of solid waste terms and the classifications vary greatly in the literature. Consequently, the use of published data requires considerable care, judgment, and common sense. The following definitions are intended to serve as a guide and are no meant to be arbitrary or precise in a scientific sense.

(1) Food wastes

Food wastes

Food wastes are the animal, fruit, or vegetable residues resulting from the handling, preparation, cooking, and eating of foods (also called garbage). The most important characteristic of these wastes is that they are highly putrescible and will decompose rapidly, especially in warm weather. Often, decomposition will lead to the development of offensive odors. In many locations, the putrescible nature of these wastes will significantly influence the design and operation of the solid waste collection system. In addition to the amounts of food wastes generated at residences, considerable amounts are generated at cafeterias and restaurants, large institutional facilities such as hospitals and prisons, and facilities associated with the marketing of foods, including wholesale and retail stores and markets.

(2) Rubbish

Rubbish consists of combustible and noncombustible solid wastes of household, institutions, commercial activities, etc, excluding food wastes or other highly putrescible material. Typically, combustible rubbish consists of materials such as paper, cardboard, plastics, textiles, rubber, leather, wood, furniture, and garden trimmings. Noncombustible rubbish consists of items such as glass, crockery, tin cans, aluminum cans, ferrous and other nonferrous metals, and dirt.

(3) Ashes and residues

Materials remaining from the burning of wood, coal, coke, and other combustible wastes in home, stores, institutions, and industrial and municipal facilities for purpose of heating, cooking, and disposing of combustible wastes are categorized as ashes and residues. Residues from power plants normally are not included in this category. Ashes and residues are normally composed of fine, powdery materials, cinders, clinkers, and small amounts of burned and partially burned materials. Glass, crockery, and various metals are also found in the residues from municipal incinerators.

(4) Demolition and construction wastes

Wastes from razed buildings and other structures are classified as demolition wastes. Wastes from the construction, remodeling, and repairing of individual residences, commercial buildings, and other structures are rubbish. The quantities produced are difficult to estimate and variable in composition, but may include dirt, stones, concrete, Brick, plaster, lumber, shingles, and plumbing, heating, and electrical parts.

(5) Special wastes

Wastes such as street sweepings, roadside litter, and litter, from municipal litter containers, catch-basin debris, dead animals, and abandoned vehicles are classified as special wastes. Because it is impossible to predict where dead animals and abandoned automobiles will be found, these wastes are often identified as originating form nonspecific diffuse sources. This is contrast to residential sources, which are also diffuse but specific in that the generation of the wastes is a recurring event.

(6) Treatment plant wastes

The solid and semisolid wastes from water and industrial waste treatment facilities are included in this classification. The specific characteristics of these materials vary, depending on the nature of the treatment process. At present, their collection is not the charge of most municipal agencies responsible for solid waste management. In the future, however, it is anticipate that that their disposal will become a major factor in any solid waste management plan.

(7) Agricultural waste

Wastes and residues resulting from diverse agricultural activities—such as the planting and harvesting of row, field, and tree and vine crops, the production of milk, the production of animal for slaughter, and the operation of feedlots—are collectively called agricultural wastes. At present, the disposal of these wastes is not the responsibility of most municipal and county solid waste management agencies. However, in many areas the disposal of animal mature has become a critical problem, especially from feedlots and dairies.

(8) Hazardous waste

Chemical, biological, flammable, explosive, or radioactive wastes that pose a substantial danger, immediately or over time, to human, plant, or animal life are classified as hazardous. Typically, these wastes occur as liquids, but thy are often found in the form of gases, solids, or sludges. In all cases, these wastes must be must be handled and disposed of with great care and caution.

Words and Expressions

zoning [ˈzəʊnɪŋ]   n. 分区，区域划分
residential [ˌrezɪˈdenʃəl]   adj. 住宅的，居住的
municipal [mjuːˈnɪsɪpəl]   adj. 市的，市政的
facility [fəˈsɪlɪtɪ]   n. 环境，设备
motel [məʊˈtel]   n. 车旅馆
fabrication [ˌfæbrɪˈkeɪʃən]   n. 生产，加工
putrescible [pjuːˈtresəbəl]   adj. 会腐败的；n. 会腐烂的物质
demolition [ˌdeməˈlɪʃən]   n. 排除，推翻
lot [lɒt]   n. 块地，块地皮
orchard [ˈɔːtʃəd]   n. 果园
dispose of   处理，处置
inclusive [ɪnˈkluːsɪv]   adj. 包括的，包含的，包含许多或一切的
encompass [ɪnˈkʌmpəs]   v. 围绕，包围，包含，包括
refuse [rɪˈfjuːz]   n. 垃圾，废物
arbitrary [ˈɑːbɪtrərɪ]   adj. 独裁的，专职的，专横的，任意的

garbge [ˈgɑːbɪdʒ]	n. （丢弃或喂猪等之）剩饭残羹垃圾
putrescible [pjuːˈtresəbəl]	adj. 腐烂的
offensive [əˈfensɪv]	adj. 人不快的，讨厌的
cafeteria [ˌkæfəˈtɪərɪə]	n. 自助食堂
institutional [ˌɪnstɪˈtjuːʃənəl]	adj. 慈善机构的
combustible [kəmˈbʌstəbl]	adj. 容易着火燃烧的
trimming [ˈtrɪmɪŋ]	n. 饰物
crockery [ˈkrɒkərɪ]	n. 陶器，瓦罐
ferrous [ˈferəs]	adj. 有铁的
cinder [ˈsɪndə]	n. 煤渣，焦渣
clinker [klɪŋkə]	n. 渣滓，熔渣，熔块
incinerator [ɪnˈsɪnəreɪtə]	n. 焚化炉
raze [reɪz]	v. 铲平，拆毁
shingle [ˈʃɪŋgəl]	n. 屋顶板
litter [ˈlɪtə]	n. 杂乱的废物
catch-basin	雨水井，沉泥井
anticipate [ænˈtɪsɪpeɪt]	v. 预料，期望
diverse [daɪˈvɜːs]	adj. 种类不同的
slaughter [ˈslɔːtə]	n. 屠宰
feedlot [ˈfiːdˌlɒt]	n. 牧场
manure [məˈnjuə]	n. 肥粪
flammable [ˈflæməbl]	adj. 易燃的
with great care and caution	以极细心和谨慎的态度
in contrast to	与……相反

Notes

1. Although any number of source classification can be developed ,the following categories have been found useful.

译文：虽然已有许多废物来源的分类法，但以下的分类法是很有价值的。

2. The most important characteristic of these wastes is that they are highly putrescible and will decompose rapidly, especially in warm weather.

译文：饮食废物的重要特点在于它具有易腐烂性，尤其在温暖天气里，会极快地腐烂。

3. In addition to the amounts of food wastes generated at residences, considerable amounts are generated at cafeterias and restaurants, large institutional facilities such as hospitals and prisons, and facilities associated with the marketing of foods, including wholesale and retail stores and markets.

译文：除了在家里产生饮食废物外，在自助食堂、饭馆、医院和监狱等机构，以及诸如批发、零售等食品市场，均会产生大量的饮食废物。

4. Typically, these wastes occur as liquids, but they are often found in the form of gases, solids, or sludges.

译文：这些废物一般呈液态，但是气态、固态及淤泥状亦较常见。

Reading Comprehension

1. What is the abstract of the test?
2. What are the sources and types of solid wastes?
3. What kinds of solid wastes do you often see? Where are they from?

Reading Material

Health Impacts of Solid Waste 固体废物对健康的影响

Modernization and progress has had its share of disadvantages and one of the main aspects of concern is the pollution affecting the earth—be it land, air, and water. With increase in the global population and the rising demand for food and other essentials, there has been a rise in the amount of waste being generated daily by each household. This waste is ultimately thrown into municipal waste collection centers from where it is collected by the area municipalities to be further thrown into the landfills and dumps. However, either due to resource crunch or inefficient infrastructure, not all of this waste gets collected and transported to the final dumpsites. If at this stage the management and disposal is improperly done, it can cause serious impacts on health and problems to the surrounding environment.

Impacts of Solid Waste on Health

The group at risk from the unscientific disposal of solid waste includes the population in areas where there is no proper waste disposal method, especially the pre-school children, waste workers, and workers in facilities producing toxic and infectious material. Other high- risk group includes population living close to a waste dump and those, whose water supply has become contaminated either due to waste dumping or leakage from landfill sites. Uncollected solid waste also increases risk of injury, and infection.

In particular, organic domestic waste poses a serious threat, since they ferment, creating conditions favorable to the survival and growth of microbial pathogens. Direct handling of solid waste can result in various types of infectious and chronic diseases with the waste workers and the rag pickers being the most vulnerable.

Exposure to hazardous waste can affect human health, children being more vulnerable to these pollutants. In fact, direct exposure can lead to diseases through chemical exposure as the release of chemical waste into the environment leads to chemical poisoning. Many studies have been carried out in various parts of the world to establish a connection between health and hazardous waste.

Waste from agriculture and industries can also cause serious health risks. Other than this, co-disposal of industrial hazardous waste with municipal waste can expose people to chemical and radioactive hazards. Uncollected solid waste can also obstruct storm water runoff, resulting in the forming of stagnant water bodies that become the breeding ground of disease. Waste dumped near a water source also causes contamination of the water body or the ground water source. Direct dumping of untreated waste in rivers, seas and lakes results in the accumulation of toxic substances in the food chain, through the plants and animals that feed on it.

Disposal of hospital and other medical waste requires special attention since this can create major health hazards. This waste generated from the hospitals, health care centers, medical laboratories, and research centers such as discarded syringe needles, bandages, swabs, plasters, and other types of infectious waste are often disposed with the regular non-infectious waste.

Waste treatment and disposal sites can also create health hazards for the neighborhood. Improperly operated incineration plants cause air pollution and improperly managed and designed landfills attract all types of insects and rodents that spread disease. Ideally these sites should be located at a safe distance from all human settlement. Landfill sites should be well lined and walled to ensure that there is no leakage into the nearby ground water sources.

Recycling too carries health risks if proper precautions are not taken. Workers working with waste containing chemical and metals may experience toxic exposure. Disposal of health-care wastes requires special attention since it can create major health hazards, such as Hepatitis B and C, through wounds caused by discarded syringes. Rag pickers and others, who are involved in scavenging in the waste dumps for items that can be recycled, may sustain injuries and come into direct contact with these infectious items.

Diseases

Certain chemicals if released untreated, e.g., cyanides, mercury, and polychlorinated biphenyls are highly toxic and exposure can lead to disease or death. Some studies have detected excesses of cancer in residents exposed to hazardous waste. Many studies have been carried out in various parts of the world to establish a connection between health and hazardous waste.

The Role of Plastics

The unhygienic use and disposal of plastics and its effects on human health have become a matter of concern. Colored plastics are harmful as their pigment contains heavy metals that are highly toxic. Some of the harmful metals found in plastics are copper, lead, chromium, cobalt, selenium, and cadmium. In most industrialized countries, color plastics have been legally banned. In India, the Government of Himachal Pradesh has banned the use of plastics and so has Ladakh district. Other states should emulate their example.

(Selected from Neil H. A.and Schubel. J. R. *Solid Waste Management and the Environment*, 1987)

History, Inheritance and Development

China Makes Notable Achievements in Environmental Protection

China has made notable achievements in protecting and improving the environment, noted a report on environmental protection law enforcement.

It was submitted to the ongoing session of the National People's Congress Standing Committee for deliberation on Tuesday, which runs from Aug. 30 to Sept. 2.

In 2021, the air quality of cities was good or excellent for 87.5 percent of the year. The number of days of heavy air pollution had fallen by 53.6 percent compared to 2015, the year the revised law on environmental protection went into effect, according to the report.

The country also enjoys clean water, read the report. The proportion of excellent and good-quality surface water reached 84.9 percent, while that of sub-standard Class V surface water dropped to 1.2 percent.

As for the prevention and control of solid waste pollution, the country enforced a ban on importing waste, promoted household waste sorting, and strengthened the supervision of hazardous waste.

Due to the efforts of law implementation, the natural ecosystems have become stable and positive, read the report. China's forest cover and stock volume have maintained growth, and the country has realized the most growth in forest resources among all countries in the world, said the report.

Efforts were also made to promote the production and use of clean energy. China has led the world in the development and utilization of renewable energy, as well as the output and sales of new energy vehicles.

The report noted the legislative efforts in environmental protection. It pointed out there are currently more than 30 laws concerning ecological and environmental protection, over 100 administrative regulations, and over 1,000 local regulations in effect.

The report also highlighted the strict law enforcement and supervision. From 2015 to 2021, China brought 175,000 prosecutions against 284,000 people suspected of crimes related to environmental and resource damage, and the courts across the country dealt with more than 977,000 environmental and resource cases of the first instance.

Practice and Training

Testing the Hardness of Water

Try this experiment with your students to measure the hardness of different water samples and investigate the effect of boiling hard water

This is a student practical, where a lot of the preparation work has been done beforehand. It could be varied so that the students watched, or carried out themselves, the preparation of the solutions. This would require using real (or simulated) sea water, rather than mixing temporarily and permanently hard water. The temporarily hard water will also really need to be boiled and cooled (as opposed to distilled water being substituted).

For younger, or less practically experienced students, consider providing the burettes already clamped and full of soap solution.

Students should bring their conical flasks to the stock bottles of solutions A to E and use a dedicated measuring cylinder for each solution to obtain 10 cm³. With larger groups, consider telling different groups to start with a different letter.

The work as described will take about 45 minutes.

Equipment Apparatus (Fig.9-2)

Eye protection

Measuring cylinders (10 cm³), x5 (one for each of the solutions A to E) Conical flask (100 cm³)

Bung, to fit the conical flask

Burette and burette stand

Small funnel

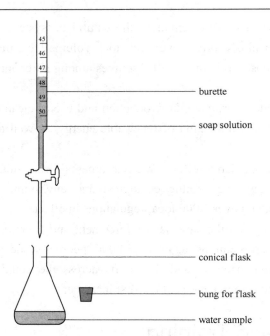

Fig.9-2 Apparatus required for testing the hardness of different water samples
Source: Royal Society of Chemistry

Chemicals

Soap solution in IDA (industrial denatured alcohol), (HIGHLY FLAMABLE, HARMFUL), 75 cm³ per group (see note 3)

A supply of distilled or deionised water for rinsing flasks between experiments Solutions as follows, about 20 cm³ per group: Solution A: deionised water, labelled as 'Rain water'; Solution B: a 50:50 mixture of temporarily and permanently hard water, labelled as 'Sea water'; Solution C: temporarily hard water, labelled as 'Temporarily hard water' (see note 4); Solution D: deionised water, labelled as 'Boiled temporarily hard water'; Solution E: permanently hard water, diluted 50:50 with deionised water and labelled as 'Boiled sea water' (see note 6)

Unit Nine Environmental Pollution and Control
环境污染及治理

Health, safety and technical notes

1. Read our standard health and safety guidance.

2. Wear eye protection throughout.

3. Soap solution in 'ethanol' (industrial denatured alcohol, IDA – see CLEAPSS Hazcard HC040A, HIGHLY FLAMMABLE, HARMFUL) can be purchased or made up. Genuine liquid soap or soap flakes, from which the liquid can be made, are increasingly difficult to obtain. Wanklyn's and Clarke's soap solutions should still be available from chemical suppliers. Lux soap flakes are ideal for making liquid soap if you can source them. Granny's Original and other non-branded soap flakes work fine but need to be used in solution as soon as they are made. They do not form a stable emulsion and precipitate out overnight. Note that most liquid hand washes are based on the same detergents as washing-up liquids and do not contain soap. To obtain soap solution from soap flakes, dissolve soap flakes (or shavings from a bar of soap) in ethanol (use IDA). See CLEAPSS Recipe Book RB000. Do not dissolve in water.

4. Dilute about 150 cm^3 of limewater (IRRITANT, see note 5) with an equal volume of distilled water. Pass in carbon dioxide (see Generating, collecting and testing gases), taking care that the gas carries over no acid spray, whereupon calcium carbonate is soon precipitated. Continue the passage of gas until all the precipitate dissolves, giving a solution of calcium hydrogen carbonate. This is temporarily hard water.

5. Limewater (calcium hydroxide solution) (IRRITANT) – see CLEAPSS Hazcard HC018 and CLEAPSS Recipe Book RB020.

6. Stir a spatula or two of hydrated calcium sulfate – see CLEAPSS Hazcard HC019B – into some deionised water. Swirl to mix, allow to stand, then decant off the clear solution. This is permanently hard water.

Procedure

1. Collect about 75 cm^3 of soap solution in a small beaker.

2. Set up a burette and, using the small funnel, fill it with soap solution.

3. Use a measuring cylinder to measure out 10 cm^3 of one of the samples of water from the list below into a conical flask: Rain water (solution A); Sea water (solution B); Temporarily hard water (solution C); Boiled temporarily hard water (solution D) Boiled sea water (solution E).

4. Read the burette. Add 1 cm^3 of soap solution to the water in the conical flask. Stopper the flask and shake it. If a lather appears that lasts for 30 seconds, stop and read the burette.

5. If no lather forms, add another 1 cm^3 of soap solution. Shake the flask. Repeat the process until a lather forms that lasts for 30 seconds. Read the burette.

6. Rinse out the flask with distilled water. Repeat the experiment with 10 cm^3 of another water sample, until you have tested them all. Make a note of the volumes of soap solution that were needed in each case to produce a lather. 7. From your experiments, decide: Which water samples are 'soft' and why; Whether sea water contains permanent hardness, temporary hardness or a mixture of both.

Teaching notes

Sample A will require very little soap solution. This shows that rain water is soft. It has effectively been distilled (and like distilled water, it will contain dissolved carbon dioxide but no salts).

Sample D will also require very little soap. This shows that temporarily hard water can be softened by boiling (see theory below).

The other samples will require more soap but E will require less than B, showing that sea water contains both temporary and permanent hardness.

The volumes of soap solution needed give a measure of the relative hardness of the various samples. With more able groups, it might be worth considering that rainwater is completely soft, so that the volume of soap required here is just the amount required to get a lather, not to overcome hardness. This volume should be subtracted from the other volumes before the relative hardnesses are compared.

Hard water contains dissolved calcium (or magnesium) salts that react with soap solution to form an insoluble scum that should be seen as a white cloudiness in the tubes:

calcium salt(aq) + sodium stearate (soap)(aq) → calcium stearate (scum)(s) + sodium salt(aq)

Only when all the calcium ions have been precipitated out as scum will the water lather. Thus the volume of soap solution measures the amount of hardness.

Temporarily hard water is defined as that which can be softened by boiling. The reactions by which it is made here are:

$$Ca(OH)_2(aq) + CO_2(g) \longrightarrow CaCO_3(s) + H_2O(l)$$

(Calcium carbonate is the 'milkiness' that forms when lime water is reacted with carbon dioxide.)

$$CaCO_3(s) + CO_2(g) + H_2O(l) \longrightarrow Ca(HCO_3)_2(aq) \text{ (calcium hydrogen carbonate)}$$

This reaction also occurs when rain water (containing dissolved carbon dioxide) flows over limestone rocks. On boiling, the reaction is reversed, softening the water:

$$Ca(HCO_3)_2(aq) \longrightarrow CaCO_3(s) + CO_2(g) + H_2O(l)$$

Permanently hard water contains calcium or magnesium salts other than the hydrogen carbonates. These are unaffected by boiling.

Additional information

This is a resource from the Practical Chemistry project, developed by the Nuffield Foundation and the Royal Society of Chemistry. This collection of over 200 practical activities demonstrates a wide range of chemical concepts and processes. Each activity contains comprehensive information for teachers and technicians, including full technical notes and step-by-step procedures. Practical Chemistry activities accompany Practical Physics and Practical Biology.

Appendix
Glossary 词汇表

A

abrasion *n.* 磨损
abrasion resistance 耐磨性
absorbent *n.* 吸附剂
absorbent *n.* 吸收剂
absorption chromatography 吸附色谱
according to 依照
accumulate *vi.* 累积；积聚
acetoacetanilide N-乙酰乙酰替苯胺，N-丁间酮酰苯胺
acetylene *n.* 乙炔
acid *n.* 酸
acrylic *n.* 丙烯酸类
active site of an enzyme 酶的活性部位
addition *n.* 加合，加入，添加
additive *n.* 添加剂
adenine *n.* 腺嘌呤
adhesion *n.* 附着
adhesive *n.* 胶黏剂 *adj.* 黏性的
adjacent *adj.* 临近的
adsorb *v.* 吸附
adsorption *n.* 吸附
aesthetic *adj.* 有关美的，审美的，悦目的
aesthetically *adv.* 审美地，美学观点上地

affinity chromatography 亲和色谱
a fractionating column 精馏塔
agglomerate *v.* 聚集
agrochemical *n.* 农用化学品，用农产品制得的化学品
airborne particulate 空中悬浮微粒
albumin *n.* 白蛋白
alcohol *n.* 醇
aldehyde *n.* 醛
aldonic acid 糖醛酸
algae *n.* 海藻，藻类（alga 的复数）
aliphatic *adj.* 脂肪族的
alkali *n.* 碱；*adj.* 碱性的
alkene *n.* 烯烃，链烯
all-too-frequent 太频繁
alter *v.* 改变
alumina *n.* 矾土，氧化铝
aluminum *n.* 铝
amber *n.* 琥珀
ambient *adj.* 周围的，环绕的；*n.* 周围环境
ambiguity *n.* 含糊，不明确
ameliorate *v.* 改善，改进
amide *n.* 酰胺

amine *n.* 胺
amino acid 氨基酸
ammonia *n.* 氨
ammonia *n.* 氨，氨水
amorphous *adj.* 无定形的，非晶的
amount *n.* 量
amphoteric *adj.* 两性的
amplify *vt.* 放大，增强
anaerobic *adj.* 厌氧的
analyte *n.* （被）分析物
anatomy *n.* 分解
ancillary *n.* 辅助
angle device 安全防护装置
angle *n.* 角
anhydrous ammonia 无水氨
anionic *adj.* 阴离子的
anionic *adj.* 阴离子的
anomalous *adj.* 异常的，不规则的
anthropogenic *adj.* 人类起源的，人为的
antibiotic *n.* 抗生素
antibody *n.* 抗体
anticipate *v.* 预料，期望
anti-clockwise 逆时针
antigen *n.* 抗原
application *n.* 应用
approach *n.* 途径，方法，手段
approach *vt.* 靠近，接近

aquatic life *n.* 水生生物
arbitrary *adj.* 独裁的，专职的，专横的
arginine *n.* 精氨酸
aromatics *n.* 芳(香)族化合物(＝aromatic compound);芳香烃
arsenic *n.* [化]砒霜，砷，三氧化二砷；*adj.* 砷的，含砷的
artificial *adj.* 人工的，人造的
artificial intelligence 人工智能
as long as 只要，在……的时候
Asbestos *n.* [矿]石棉；*adj.* 石棉的
asbestos *n.* 石棉
a series of 一系列
asphalt *n.* 沥青
asphyxia *n.* 窒息
asphyxiate *vt.* 使……窒息
assay *n.* 分析
associated gas 与石油同时发现的伴生气（天然气）
assumption *n.* 假定
atmosphere *n.* 气氛，大气，空气
atmospheric pressure 气压
attainment *n.* 开车率，开工率
a typical chemical process 典型化工生产工程
average *n.* 平均数
axial *adj.* 轴的，轴向的，轴心的
azide *n.* [化]叠氮化物

B

base *n.* 碱基
batch processes 间隙生产过程
bauxite *n.* 铝土岩（产铝的矿土、石）
beaker *n.* 高脚杯；烧杯

befall *vt.* 落到……的身上，降临于
bentonite *n.* 斑脱土（火山灰风化的胶状黏土），膨润土，皂土
benzene *n.* 苯

bioactivity *n.* 生物活性
biocatalysis *n.* 生物催化
biochemical engineering 生物化工，生化工程
biological synthesis 生物合成
biopond *n.* 生物池（生物废水处理装置）
bioprocess *n.* 生物工艺
bioreactor *n.* 生物反应器
biosphere *n.* 生物圈
biosynthesis *n.* 生物合成
biotechnology *n.* 生物技术
blade *n.* 刀锋，刀口
boil *v.* 沸腾

boiler feed water 锅炉给水
boiling point 沸点
bolt *n.* 螺栓
bond-breaking reactions 键断裂反应
bracket *v.* 包括，囊括
branched *adj.* 支化的
break down 打破，分解
broth *n.* 肉汤（培养基）
bullet-proof vest 防弹衣
buret *n.* 滴定管，玻璃量管
butadiene *n.* 丁二烯
butane *n.* [化]丁烷
by-product *n.* 副产品

C

cadmium *n.* 镉（Cd）
cafeteria *n.* 食堂
calcium *n.* 钙（Ca）
calculation *n.* 计算
calculator *n.* 计算器
calorific value [化]热值；发热量
capillary *n.* 毛细管；*adj.* 毛状的，毛细作用
carbohydrate *n.* 糖类，碳水化合物
carbohydrate *n.* 碳水化合物，糖类
carbon dioxide 二氧化碳
carbon *n.* 碳
carboxyl group 羧基
carboxylate *n.* 羧酸盐，羧酸酯
carcinogen *n.* 致癌物（质），诱癌因素
carcinogenic *adj.* 致癌物的，致癌的
cargo *n.* 货物，一批（货物）
carrier gas 载气
cast iron 铸铁

casualty *n.* 事故，灾难，死伤
catalyst life 催化剂寿命
catalyst *n.* 催化剂
catalyst particles 催化剂颗粒
catalytic *adj.* 催化的
catalytic cracking 催化裂化
catalyze *v.* 催化
catastrophic *adj.* 悲惨的，灾难的
catch-basin *n.* 雨水井，沉泥井
category *n.* 种类
cation *n.* 阳离子
cationic *adj.* 阳离子的
caustic *adj.* 腐蚀性的
CE marked CE 标志认证
cell *n.* 细胞
cellular *adj.* 细胞的
cellulose *n.* 纤维素
cement *n.* 水泥

centrifugal compressor 离心式压缩机
ceramic *n.* 陶瓷，*adj.* 陶瓷的
chemical engineering 化学工程
chemical engineering principle 化工原理
chloride *n.* 氯化物
chlorine *n.* 氯
chloroformates *n.* 氯甲酸酯
chlorthalidone *n.* [药]氯噻酮（利尿降压药）
cholera *n.* 霍乱
cholinesterase *n.* 胆碱酯酶
chromatographic *adj.* 色谱的
chromatographic column 柱效
chromatography *n.* 色谱，套色板，层析
cinder *n.* 煤渣，焦渣
circular saw 圆锯
circulation *n.* 流通，循环
clarify *v.* 使……澄清
clarity *n.* 透明性
clinker *n.* 渣滓，熔渣，熔块
closed-loop 闭环
clot *v.* 凝结
coalescence *n.* 聚集
coalescer *n.* 聚结器
coastline *n.* 海岸线
code *n.* 法律，法规
coefficient *n.* 系数
cohesive energy 内聚能
coil *v.* 盘绕
coin *vt.* 制造（字句），杜撰
collide *vi.* or *vt.* 碰撞，互撞
colloidal *adj.* 胶体的
colorfast *adj.* 不褪色的
colorless *adj.* 无趣味的，苍白的，无色的

column *n.* 柱
combustible *adj.* 容易着火燃烧的
combustion *n.* 燃烧，氧化
commitment *n.* 承诺，应允的义务
commodity *n.* 日用品
commuter *n.* 乘公交车上下班者
compact *adj.* 紧凑的
compensation *n.* 补偿，赔偿，酬报
component *n.* 组成
composite *adj.* 合成的，复合的；*n.* 合成物，复合材料；*vt.* 合成
compound *n.* 配合，配混
compressor *n.* 压气机，压缩机
compute *v.* 计算
concentration *n.* 浓度
concentration *n.* 浓缩
concurrent *adj.* 并流的，顺流的
condensation *n.* 压缩，凝结，冷凝
conduction *n.* 传导
configuration *n.* 构造，结构，构型，构象
consensus *n.* 一致
constant *n.* 常数
constituent *n.* 成分
consume *v.* 消耗
contact *n.* 接触
contaminant *n.* 污染物；致污物
content *n.* 含量
continuous processes 连续生产过程
control panel 控制面板
convection *n.* 对流
convene *vt.* 召集，集合
conversion *n.* 转化，转换
convert *v.* 转化，转变

cool *v.* 冷却
coolant *n.* 冷冻剂
copper *n.* 铜
copy *v.* 复制
Cornell University 美国康奈尔大学
cornerstone *n.* 基础
corrosion *n.* 腐蚀，腐蚀产生的物质，衰败
corrosive *adj.* 腐蚀的，侵蚀性的；*n.* 腐蚀物
cosmetic *adj.* 美容的，化妆用的；*n.* 化妆品，装饰品
craft *n.* 手艺，技艺
crane *n.* 起重机
crazy *adj.* 疯狂的，发狂的

criterion *n.* 标准
critical *adj.* 批评的，严谨的
critical velocity 临界速度
crockery *n.* 陶器 瓦罐
crosslinking *n.* 交联
cryogenic *adj.* 低温的
cryptosporidium *n.* 隐孢子虫
crystalline *n.* 结晶
culture *n.* 培养
cumbersome *adj.* 麻烦的，笨重的
cumulative *adj.* 累积的，累积性的
cyanide *n.* 氰化物
cyclone *n.* 旋风；旋风分离器
cytosine *n.* 胞嘧啶

D

damp *v.* 阻尼
dealkylation *n.* 脱烷基化作用
Declaration of Conformity 符合性声明
defection *n.* 缺乏
defence *n.* 防御
defensive *adj.* 防御的，防卫的
degradation *n.* 降解
degrade *v.*（使）降解
deheptanizer column 脱庚烷塔
dehydrogenation *n.* 脱氢作用
delineate *vt.* 叙述，描写
demolition *n.* 排除，推翻
denature *vt.* 使变性
density *n.* 密度
deoxyribofuranose *n.* 脱氧核糖呋喃糖
deoxyribose *n.* 脱氧核糖
deoxyribose nucleic acid 脱氧核糖核酸

depict *vt.* 描绘，描画，描述
deplete *v.* 使减小
depletion *n.* 消耗，损耗
deployment *n.* 使用，利用，推广应用
depolymerize *v.*（使）解聚
derivative *n.* 衍生物
desalination *n.* 脱盐
desired reaction 目标反应
despoil *vt.* 抢劫，掠夺
detergent *n.* 清洁剂，去垢剂
detrimental *adj.* 有害的，不利的
dexterity *n.*（手）灵巧，熟练
diagram *n.* 图
dielectric *n.* 电介质
diethanolamine *n.* 二乙醇胺
differential equation 微分方程
differentiate *v.* 区别，区分

diffudivity *n.* 扩散性，扩散系数
diffusivity *n.* 扩散性
dilatometry *n.* 膨胀法
dilute *adj.* 稀的；*v.* 稀释
dinitrobenzene *n.* 二硝基苯
dismantle *v.* 拆卸；拆除……的设备；摧毁
displace *v.* 置换
disposal *n.* 处置，处理，处理方式
disposal *n.* 清理，处理
dispose of 处理，处置
dissipation *n.* 损耗
dissolve *v.* 溶解
distillation *n.* 蒸馏，蒸馏法

diuretic *adj.* 利尿的
diverse *adj.* 种类不同的
dizziness *n.* 头晕，头昏眼花
double bond 双键
downstream *adj.* 下游的
drain *n.* 排水沟，阴沟
drilling machine 钻孔机
drying *n.* 干燥；*adj.* 烘干的
durability *n.* 持久性，耐久性
dyad *n.* 二联体
dye *n.* 染料；*v.* 染色
dysentery *n.* 痢疾

E

economically *adv.* 经济的
ecosystem *n.* 生态系统
eddy *n.* 漩涡
effluent *n.* 流出物，污水，废气
ejector pins *n.* 顶杆
elasticity *n.* 弹性
elastomer *n.* 弹性体
electronics *n.* 电子学（单数）
electrostatic *adj.* 静电的，静电学的
ellipse *n.* 椭圆
elute *vt.* （化）洗提
elution *n.* 洗脱
embed *vt.* 把……嵌入，埋入
emergence *n.* 出现，浮现
empiricism *n.* 经验主义
emulsifier *n.* 乳化剂
enamel *n.* 搪瓷，珐琅，釉药，瓷漆
enclose *vt.* 密封，围包住

encompass *v.* 围绕，包围，包含，包括
end point 终点
endothermic *adj.* 吸热的，吸能的
Energy Information Administration 能源信息管理局
enhance *v.* 促进，提高
enol *n.* 烯醇式
ensue *vi.* 跟着发生，继起
entail *v.* 使必需，使承担
entanglement *n.* 纠缠，缠结
entity *n.* 实体
enzyme *n.* 酶
epidemic *n. adj.* 流行（的），传染（的）
equilibrium *n.* 平衡
equipment *n.* 设备
erlenmeyer flask 锥形烧瓶，爱伦美氏（烧）瓶
erosive *adj.* 腐蚀的，侵蚀性的

escalate *vt.* 使逐步升级（加剧）
essential materials 原料
ether *n.* 醚
EU= European Union 欧洲联盟（简称欧盟）
evacuation *n.* 撤退，撤散，疏散
evaporate *vt.* 使……蒸发，使……脱水
evaporation *n.* 蒸发
eventually *adv.* 最终
evolve *vt.* 发展，使演变
exhaustive *adj.* 耗尽的，枯竭的
exothermic *adj.* 放热的
experimentally *adv.* 实验地，通过实验地
explode *vt.* 使爆炸，破除，戳穿
expression system 蛋白质表达系统
extracellular *adj.* 细胞外的
extraction *n.* 提取

F

fabrication *n.* 生产，加工
facilitate *vt.* 促进，帮助
facility *n.* 环境，设备
fahrenheit *n.* 华氏温度计，华氏温标
fat *n.* 脂肪
fatality *n.* 灾祸
feature *n.* 性质
feedlot *n.* 牧场
feedstock *n.* （工业加工用的）原料，尤指（用于制造石油化学产品的）化工物
feedstock *n.* 原料
ferric *adj.* 铁的，含铁的，（正）铁的，三价铁的
ferrous *adj.* 有铁的
fiber *n.* 纤维
fin *n.* 翅片
fine chemicals 精细化学品
fire retardance 阻燃性
firefighting *n.* 消防
fission *n.* （原子的）分裂，裂变
fixed-bed reactor 固定床反应器
flammable *adj.* 易燃的
flaw *n.* 裂纹，有瑕疵
flax *n.* 亚麻，麻布，亚麻织品
flexibility *n.* 柔软性
flow diagram 流程图
fluctuation *n.* 波动，起伏
fluid *n.* 流体
fluidity *n.* 流体状态
folding *n.* 折叠
foodstuff *n.* 食品
fork lift truck 叉车
formation *n.* 形成，生成
formula *n.* 公式，规则
fouling *n.* 沉积物
fraction *n.* 分部，部分
fraction *n.* 馏（部）分
fractionation *n.* 分馏法
fragment *n.* 链段
frailty *n.* 意志薄弱，性格缺陷
fuel cells 燃料电池
fuller's earth 漂白土，埃洛石，高岭石，蒙脱石
fume *n.* (pl.) 烟气（雾），浓烟
functional group 官能团
fundamental *adj.* 基本的
funnel *n.* 漏斗

fused silica 熔融石英

fusion *n.* 熔合，合并热核反应

G

garbge *n.* 剩饭残羹，垃圾
gas-phase reactor 气相反应器
gear *n.* 装置，变速机构
gel chromatography 凝胶色谱
gel filtration 凝胶过滤（即凝胶色谱）
genetic *adj.* 遗传的
geometric *adj.* 几何的
girder *n.* 主梁，大梁
glacier *n.* 冰河，冰川
glossy *adj.* 光洁的，光滑的
glucagon *n.* 胰高血糖素

glucose *n.* 葡萄糖
glycine *n.* 甘氨酸
glycogen *n.* 糖原
graded *adj.* 分级的，递级的
gradient *n.* 梯度
greenhouse gas（GHG）温室气体
grinding machine 磨床
guanine *n.* 鸟嘌呤
guillotine *n.* 剪床
gypsum *n.* 石膏

H

hand saw 手锯
helical *adj.* 螺旋的
helium *n.* 氦
hematopoietic *adj.* 造血的，生血的
hemoglobin *n.* [生化]血红素，[生化]血红蛋白（等于 haemoglobin）
heparin *n.* 肝素钠
hepatitis *n.* 肠炎
herbicide *n.* 除草剂
heritage *n.* 继承物，遗产
hermetically *adv.* 密封地
homolytic *adj.* 均裂的
hormone *n.* 激素
HSE 健康、安全、环境管理体系的简称
hybridization *n.* 杂交，杂化

hydraulic *adj.* 水力的，水压的
hydrocarbon *n.* 碳氢化合物
hydrocarbon *n.* 碳氢化合物
hydrochloric acid 盐酸
hydrocracker 加氢裂化装置
hydrodesulfurization *n.* 加氢脱硫（过程），氢化脱硫作用
hydrogen fluoride *n.* 氟化氢
hydrogen sulfide *n.* 硫化氢
hydrogen-bond 氢键
hydrolysis *n.* 水解
hydrosilicate *n.* 含水硅酸盐
hydroxyl *n.* [化]羟（基），氢氧基，氢氧化物

I

immune *adj.* 免疫的，不受影响的

immune affinity chromatograph 免疫亲和色谱

immunoaffinity *adj.* 免疫亲和性

immunological *adj.* 免疫学的

impeller *n.* 推进者，轮叶

imperative *adj.* 必要的，急迫的

implant *v.* 植入

implementation *n.* 贯彻，实施，手段

impurity *n.* 不纯，杂质

in contrast to 与……相反

in situ 在原位置，在原处

in vivo 在活的有机体内

incinerator *n.* 焚化炉

inclusive *adj.* 包括的，包含许多或一切的

indicator *n.* 指针，指示器，记录器指示物，指示者

inert *n.* 惰性组分

inert gas 惰性气体

inertness *n.* 不活泼，惰性

influenza *n.* 流行性感冒（简写 flu）

ingenuity *n.* 机灵，独创性，精巧

inhalation *n.* 吸入

inhibit *v.* 抑制

initial *adj.* 起始的，最初的

initiator *n.* 引发剂

inoculums *n.* 接种液

institutional *adj.* 慈善机构的

institutionalize *vt.* 使制度化

insulation *n.* 绝缘

insurance premium 保险费

interaction *n.* 相互作用

interconversion *n.* 相互转化

interdisciplinary *adj.* 多学科交叉的

interferon *n.* 干扰素

interleukin *n.* 白介素

intermolecular *adj.* 分子间的

internal *adj.* 内部的

interplay *v.* 相互作用

interpolate *v.* 内插，在两个已知值之间估计（函数或级数）的一个值

interpose *vt.* 放入，插入

intracellular *adj.* 细胞内的

intrarmolecular *adj.* 分子内的

inventory *n.* 库存

invert *v.* 上下颠倒

investigation *n.* 调查，调查研究

ion-exchange chromatography 离子交换色谱

ironing *n.* 熨平

irreversible physical change 不可逆物理变化

irritant *n.* 刺激性物质，刺激性气体

irrotational flow 无旋流

isoindolinyl *n.* 异二氢氮杂茚基

isolating valve 隔离阀

Isomar UOP 公司注册商标

isomerization *n.* 异构化，异构化作用

isoprene *n.* 异戊二烯

italics *n.* 斜体字

K

keto *n.* 酮式

ketone *n.* 酮

kinetics *n.* 动力学

L

laboratory *n.* 实验室
labour-intensive 劳动强度大的
large scale industrial equipment 大型工业装置
lathe *n.* 车床
launch *v.* 使运动、送上轨道
laureate *adj.* 佩戴桂冠的；*n.* 戴桂冠的人
legitimacy *n.* 合法性
lethal *adj.* 致命的，致死的
leveraged *n.* （达到某目的的）手段，力量，优势，影响力，作用力
lid *n.* 盖子
lifting sling 升降索套
light ends 轻馏分
lignin *n.* 木质素
limestone *n.* 石灰石
linear *adj.* 线的，线性的
linkage *n.* 连接
lipid *n.* 脂
liquid *n.* 液体
liquid-propellant 火箭引擎中的液体燃料
list *v.* 排列
litter *n.* 杂乱的废物
liver *n.* 肝脏
live-stock 家畜，牲畜
Lloyd's Register 劳埃德船级社
logarithm *n.* 对数
lot *n.* 块地，块地皮
lubricant *n.* 润滑物，润滑油，润滑剂
lubricating oil *n.* 润滑油
lung *n.* 肺，呼吸器
lyophobic *adj.* 疏水的

M

macromolecule *n.* 高分子
magnesium *n.* 镁（Mg）
magnetic stirrer 磁搅拌器，磁力搅拌机，磁性搅拌器
magnitude *n.* 大小，数量，量级
makeup *n.* 组成；补充
malleable *adj.* 有延展性的
malonate *n.* 丙二酸
manure *n.* 肥粪
mechanical *adj.* 机械的
mechanical properties 力学性能，机械性能
median *adj.* 平均的
mega *n.* 百万，大
membrane *n.* 膜
meniscus *n.* 新月，半月板
mercaptan *n.* [化]硫醇
mercury thermometer 水银温度计
metabolic *adj.* 代谢的
metabolic pathway 代谢途径
metabolite *n.* 代谢产物
methacrylate *n.* 甲基丙烯酸酯
methane *n.* 甲烷，沼气
methemoglobinemia *n.* 高铁血红蛋白症
methionine *n.* [生化]蛋氨酸，甲硫氨酸

methyl isocyanate 甲基异氰酸盐
meticulously *adv.* 精心地,精确地
mica *n.* 云母
microbial *adj.* 微生物的
microorganism *n.* 微生物,微小动植物
migrate *vi.* 迁移
milling machine 铣床
Mil-Std-882 军标-882
minor *adj.* 较小的,较次要的
mishap *n.* 灾祸
mitigate *vt.* 减轻;使缓和
mixer *n.* 搅拌器
MMBtu 代表百万英热单位,百万英制热单位
mobile phase 流动相
mobile source 流动污染源

model *v.* 建模
modification *n.* 修饰,改进
mole *n.* 摩尔
molecular *adj.* 分子的
molecular packing 分子排列
molecular sieves 分子筛
molecular size 分子大小
molecular-weigh 分子量
monitor *v.* 监控
monoethanolamine *n.* 单乙醇胺
monomer *n.* 单体
motel *n.* 汽车旅馆
mowing machine 割草机
mucous *adj.* 黏液的,黏液似的
multi tubular reactor 管式反应器
municipal *adj.* 市政的

N

namely *adv.* 即,那就是
naphtha *n.* 石脑油
naphthalene *n.* 萘
naphthene *n.* 环烷烃
NASA 美国国家航空航天局
necessitate *vt.* 使成为必需
neglect *vt.* 忽视
negligible *adj.* 无关紧要的

neoprene *n.* 氯丁橡胶
net production 净产量,净生产量
nitrogen oxide *n.* 氮氧化物
nucleotide *n.* 核苷酸
number *v.* 编号
nutrient *n.* 营养成分
nylon *n.* 尼龙

O

octane *n.* 辛烷
odorless *adj.* 没有气味的
offensive *adj.* 令人不快的,讨厌的
olefin *n.* [化]烯烃
onion *n.* 笨蛋;搞糟的事情

operation *n.* 操作
optical *adj.* 光学的
optimization *n.* 优化
orchard *n.* 果园
organic compound 有机化合物

orientation *n.* 取向
OSHA 职业安全与健康管理总署
outfall *n.* 出口，排口，河口

oxidation and reduction 氧化和还原
oxidation *n.* 氧化作用

P

palladium *n.* 钯
parachute *n.* 降落伞
paracrystalline *n.* 先结晶
paradigm *n.* 范例
paraffin *n.* 石蜡，[化]链烷烃
parameter *n.* 参数；参量；系数
parasite *n.* 食客，寄生虫
partial *adj.* 部分的，热爱的
particle *n.* 颗粒
particular *adj.* 特定的
particulate matter *n.* 微粒物质，悬浮微粒
patent *v.* 取得专利权 *n.* 专利，许可证
pathogen *n.* 病原体，病菌
peptide bond 肽键
peptide *n.* 多肽
per pass 单程
percutaneous *adj.* 经由皮肤的
peril *n.* 危险，冒险
permanently *adv.* 永久地
peroxide *n.* 过氧化物
peroxyacyl nitrates 过氧乙酰硝酸酯
perpendicular *adj.* 垂直的
Persian Gulf 波斯湾
pesticide *n.* 杀虫剂，农药
petrify *n.* 吓呆，使麻木
petroleum *n.* 石油
petroleum refinery 炼油厂
pharmaceutical *adj.* 制药（学）上的；*n.* 药剂，药物
phenomena *n.* 现象（phenomenon 的复数）
phosphate diester group 磷酸二酯基团
phosphate groups 磷酸基团
phosphorus *n.* 磷
photochemical oxidant [化]光化学氧化剂
photocopier *n.* 复印机
phthalate *n.* 邻苯二甲酸盐
physiology *n.* 生理
picoline *n.* [化]皮考啉，甲基吡啶
pipet *n.* 吸量管，球管
plane *n.* 平面
plant *n.* 车间，工厂
plasma *n.* 血浆
plastic *n.* 塑料
plasticizer *n.* 增塑剂
platinum *n.* 铂，白金（符号为 Pt）
plausible *adj.* 稳定的
polio *n.* 脊髓灰质炎（即 poliomyelitis）
polyester *n.* 聚酯
polyethylene *n.* 聚乙烯
polymer *n.* 聚合物
polymeric *adj.* 聚合的，聚合物的
polymerism *n.* 聚合现象
polyoxyethylene *n.* 聚乙氧基
polysaccharide *n.* 多糖，聚糖，多聚糖
porcelain *n.* 瓷器，瓷
portable *adj.* 便携的，轻便的

Glossary 词汇表

portfolio n. 业务量，业务责任
position n. 位置
post-translational modification 翻译后修饰
postulate v. 假设
potassium n. 钾（K）
potassium thiocyanate 硫氰酸钾
potency n. 效价
pozzolanic adj. 凝硬性的，火山灰的
ppmv 按体积计算百万分之一
predictable adj. 可预测的
preference n. 偏爱
preliminary adj. 预备的，初步的
preliminary adj. 预先的
preoccupation n. 偏见，成见
preparation n. 制备
pressure vessel 压力容器
pressurization n. 加压
prevalent adj. 流行的，普遍的
primary-standard 基准物
principle n. 原理，法则，原则
process n. 过程，加工，操作

process v. 处理
procurement contracts 采购合同
product n. 产物，产品
propagation n. 传播
propane n. [化]丙烷
property n. 性能
protein n. 蛋白质
proteinaceous adj. 蛋白质的，似蛋白质的
prototype n. 主型，原型
protozoa n. 原生动物，原生动物（protozoan 的复数）
proven reserves 探明储量
pulmonary adj. 肺部的
purification n. 净化，提纯
purine n. 嘌呤
putescible adj. 腐烂的
putrescible adj. 会腐败的；n. 会腐烂的物质
pyrimidine n. 嘧啶
pyrolysis n. 高温分解，
pyrolysis gasoline 热解汽油
pyrophoric adj. 发生火花的，生火花的

Q

quantitative analysis 定量分析
quantitatively adv. 定量地

quantity n. 量
quench tower 急冷塔

R

radioactive cloud 放射云
rate n. 速率
ratio n. 比例
rational adj. 合理的
rationale n. 基本原理
rationalization n. 合理化

rayon n. 人造丝纤维
raze v. 铲平，拆毁
reactant n. 反应物
reactor effluent 反应器流出物
reagent n. 试剂
reagent n. 反应物，试剂

recombinant *adj.* 重组的
reduction *n.* 还原
refinery *n.* 精炼厂，提炼厂，冶炼厂
reflux *n.* 回流
reformate *n.* 重整油，重整产品
refractive *adj.* 折射的
refractometer *n.* 折射计
refuse *n.* 垃圾，废物
regenerate *v.* 再生
relevant *adj.* 信赖的，依靠的
removal *n.* 消去
repel *vt.* 排斥，相斥
replication *n.* 重复试验
residential *adj.* 住宅的，居住的
residual *adj.* 残余的
residue *n.* 残基

resilient *n.* 弹性，有弹性；*adj.* 反弹的
resin *n.* 树脂
respiratory *adj.* 呼吸的，呼吸系统的
respiratory tract 呼吸道
retention factor 保留因子
retention time 保留时间
revenue *n.* 税收，收益，总收入
reverse micellar 反胶团
rhenium *n.* 铼（符号为 Re）
ribose *n.* 核糖
robust *adj.* 加强的，增强的
ROPS 翻车安全保护装置
rotate *v.* 旋转，转动
rule-of-thumb 单凭经验的方法
runoff *n.* 径流
rupture *v.* 破裂

S

safety device 安全防护装置
saline *n.* 生理盐水
saturate *v.* 饱和
saturated *adj.* 饱和的
scale down 降低，减小
scratch *n.* 抓，抓痕，刮擦声
screwdriver *n.* 螺丝起子
scum *n.* 泡沫，浮渣
secrete *v.* 分泌
seizure *n.* 抓住，攫取，夺取
selection *n.* 选择
selective *adj.* 选择性的
selectivity factor 选择性因子
separation *n.* 分离
septic tank 化粪池

sequence *n.* 序列，顺序
shale gas 页岩气
shedding *n.* 脱落，流出，散发
shigle *n.* 屋顶板
shovele *v.* 铲起，把……大量倒入
shutdown *n.* 关闭，倒闭，关机，停工
sieving *n.* 筛分
silica *n.* 硅土，二氧化硅
siloxane *n.* 硅氧烷
silver nitrate 硝酸银
size reduction 粉碎
slaughter *n.* 屠宰
small laboratory apparatus 小型实验装置
sodium chloride 氯化钠
sodium hydroxide 氢氧化钠

sodium *n.* 钠（Na）
sodium *n.* 钠盐
solid *n.* 固体
solubility *n.* 溶解度
solution *n.* 溶液
solvent *n.* 溶剂
space velocity 空速
specific gravity 密度
spewing debris 压榨碎片
spherulites *n.* 球晶
spur *vt.* 加速，鞭策；*vi.* 急速前进
stabilizer 稳定塔
standard *n.* 标准品
standard solution 标准溶液
standardization *n.* 标准化，标定
stationary phase 固定相
stationary source 固定污染源，固定来源
sterilize *vt.* 消毒，杀菌
stiffness *n.* 刚性
stimulate *v.* 刺激
stoichiometry *n.* 化学计算（法）；化学计量学
stopcock *n.* 管闩，活塞，活栓，旋塞阀
streptomycin *n.* 链霉素
stripper *n.* 汽提塔
subsequently *adv.* 随后
substitution *n.* 取代，代替
substrate *n.*（供绘画、印刷等的）底面，基底，基片；[生]酶作用物（有机体的）培养基

substrate *n.* 底物
subtract *vt. & vi.* 减，扣除，做减法
subunit *n.* 亚基
successive *adj.* 连续的
suction *n.* 吸，抽吸
sufficient *adj.* 足量的
sugar *n.* 糖
suggest *v.* 提出
sulfate *n.* 硫酸盐
sulfate *n.* 硫酸盐；*vt.* 使成硫酸盐，用硫酸处理
sulfonamide *n.* 磺胺
sulfonate *n.* 磺酸盐
sulfur dioxide 二氧化硫
sulfur *n.* 硫
sulfuric acid 硫酸
sulphate *n.* 硫酸盐
superheat *vt.* 使……过热；*n.* 过热
superintendent *n.* 监督人，指挥者
supernatant *n.* 上清液
surfactant *n.* 表面活性剂；*adj.* 表面活性剂的
surgical *adj.* 外科的；*n.* 外科手术
surveillance *n.* 监测，监督
suspended *adj.* 悬挂着的，吊着的
swamp gas 沼气（等于 smarsh gas）
swarf *n.* 切屑
symmetry *n.* 对称
synonymous *adj.* 同义词的

T

tablet *v.* 把……压成片（块），制片（块）
tabulate *v.* 数据列表

tautomeric *adj.* 互变异构的
TBC 对叔丁基邻苯二酚

Teflon *n.* 特氟纶,聚四氟乙烯
tensile strength 拉伸强度
terminology *n.* 术语，术语学，用辞
test tube 试管
textile *n.* 纺织品，织物
theoretical mathematical prototypes 理论数学模型
therapeutic *adj.* 治疗的
therapeutically *adv.* 治疗上
thermal cracking 热裂解
thermodynamic *adj.* 热力学的
thermodynamics *n.* 热力学
thermoset *adj.* 热固性的 *n.* 热固性
three-dimensional network 三维网状结构
thrombin *n.* 凝血酶
thymine *n.* 胸腺嘧啶
tight oil 致密油
tilt *v.* 倾斜
time line 年表，活动时间表
time *v.* 计时
tissue plasinogen activator 组织纤溶酶原激活剂

titrant *n.* 滴定剂，滴定（用）标准液
titration error 滴定误差
titration *n.* 滴定
titrimetric *adj.* 滴定（测量）的
toluene *n.* 甲苯[亦称作 methylbenzene, phenylmethane]
tough *adj.* 韧性的
toughness *n.* 韧性
trial *n.* 试验，审判
triazine 三嗪
trigger *vt.* 出发，激发
trim *vt.* 调整
trimethylammonium *n.* 三甲基铵化物
trimming *n.* 饰物
tubular *adj.* 管状的
tubular reactor 管式反应器
turbidimeter *n.* 浊度计，浊度表
turbidity *n.* 混浊，混乱
turbulence *n.* 湍流
turbulent *adj.* 狂暴的，湍急的
typeface *n.* 字体，铅字字面
typhoid *adj.* 伤寒的；*n.* 伤寒

U

undergo *v.* 经历
underneath *prep.* 在……下面
undisturbed *adj.* 不受打扰的
unit *n.* 单位（活力单位）

unity *n.* 单位
unsightly *adj.* 难看的，不雅观的
upstream *adj.* 上游的
utilize *v.* 利用

V

vacuum pedestal 真空助力器座
valve *n.* 阀
van der Waals 范德华

vapor *n.* 蒸汽，烟雾
variable *n.* 变量，可变参数
vehicle hoist 升车机

velocity *n.* 速度，速率

vent *n.* 出口，孔

vice versa 反之亦然

vicinity *n.* 邻近，附近，近处

vinegar *n.* 醋

viruses *n.* 病毒（virus 的复数）

viscosity *n.* 黏度

vital spirit 生命的精气

vitamin *n.* 维生素

vitreous *adj.* 玻璃（似）的，玻璃质的

volatile *adj.* 挥发（性）的

volume *n.* 体积

volumetric flask （容）量瓶

VTL（vertical turret lathe） 立式转塔车床

vulcanize *v.* 硫化

W

wafer *n.* 晶片，圆片

wearability *n.* 耐磨性，磨损性

weathering resistance 耐候性

weight *v.* 称重

welding *n.* 焊接

windblown *adj.* 被风吹的，风飘型的

with great care and caution 以极细心和谨慎的态度

workpiece *n.* 工件

worm *n.* 蛆，寄生虫

wrap *v.* 包，裹，卷

X

xylan *n.* 木聚糖

xylene *n.* [化]二甲苯

Y

yield point 屈服点

Z

z-coordinate z 轴坐标

zeolite *n.* 沸石

zwitterionics *adj.* 两性离子的

zoning *n.* 分区，区域划分

Reference　参考文献

[1] R. Gani, E. A. Brignole. Molecular design of solvents for liquid extraction based on UN IFAC [J]. Fluid Phase Equilibria, 1983, 13 (1):331-340.

[2] E.A. Brignole, S. Bottini, R. Gaul. A strategy for the design and selection of solvents for separation processes [J]. Fluid Phase Equilibria, 1986, 29 (1) :125 - 132.

[3] V. Venkatasubramanian, K. Chart, J. M. Caruthers. Computer-aided molecular design using genetic algorithms [J]. Computers & Chemical Engineering, 18 (9): 833 - 844.

[4] Braam van Dyk, Izak Nieuwoudt. Design of solvents for extractive distillation [J]. Iud Eng Chem Res, 2000, 39 (5): 1432 - 1429.

[5] James G. Speight. Chemical Process and Design Handbook[MI. MCGRAW-HILL, 2002.

[6] Douglas A. Skoog, F. James Holler, Stanley R. Crouch Fundamentals of Analytical Chemistry[M]. 8 thed. Cengage Learning,2009.

[7] David Almorza Gomar, C.A. Brebbia, D. Almorza, D. Sales. Solid Waste Management and the Environment[M]. WIT Press ,2002.

[8] Daniel D. Chiras. Environmental Science[M]. 8th ed. New York: The Drylen Press, 1984.

[9] 刘琼琼．高分子材料专业英语[M]．北京：化学工业出版社，2005．

[10] 揣成智．高分子材料工程专业英语[M]．北京：中国轻工业出版社，1999．

[11] 曹同玉，冯连芳．高分子材料工程专业英语[M]．北京：化学工业出版社，2008．

[12] 张晓黎，李海梅．塑料加工和模具专业英语[M]．北京：化学工业出版社，2005．

[13] 程为庄，顾国芳．大学专业英语阅读教程：高分子材料[M]．上海：同济大学出版社，1999．

[14] 司鹄．安全工程专业英语[M]．北京：机械工业出版社，2007．

[15] 樊运晓．安全工程专业英语[M]．北京：化学工业出版社，2006．

[16] 李居参．工业分析专业英语[M]．北京：化学工业出版社，2004．

[17] 姜彦，刘晓兰．生物工程专业英语[M]．哈尔滨：哈尔滨工程大学出版社，2006．

[18] 许赣荣．发酵生物技术专业英语[M]．北京：中国轻工业出版社，2007．

[19] 吴昊，乔德阳．生物化工与制药专业英语[M]．北京：化学工业出版社，2009．

[20] 羌宁．环境工程—：专业英语[M]．北京：化学工业出版社，2004．

[21] 宋志伟．环境专业英语教程[M]．哈尔滨：哈尔滨工业大学出版社，2005．

[22] 钟理．环境专业英语[M]．北京：化学工业出版社，2005．

[23] 李居参．环境专业英语[M]．北京：化学工业出版社，2007．

[24] 王子康，雪文，杨晋．专业英语(石油加工)[M]．北京：中国石化出版社，1991．

[25] 胡鸣，刘霞．化学工程与工艺专业英语[M]．北京：化学工业出版社，2008．

[26] 波任 M．E.，等．无机盐工艺学[M]．北京：化学工业出版社，1995．

[27] 从丛，李咏燕．学术交流英语教程[M]．南京：南京大学出版社，2003．

[28] Wanhua Chemical attended the PU CHINA 2021 with its life-cycle sustainable solutions for polyurethane materials [EB/OL]. https://en.whchem.com/cmscontent/606.html

[29] China unveils plan to strengthen safe production of hazardous chemicals [EB/OL]. https://www.chinadaily.com.cn/a/202203/21/WS62382915a310fd2b29e52484.html

[30] Scientists find cheap way to turn gases into liquids [EB/OL]. https://usa.chinadaily.com.cn/a/201808/23/WS5b7e1e82a310add14f387517.html

[31] 世界技能大赛官网简介 https://worldskills.org/

[32] ROUT, Deeleep, Kumar ;SINHA, Ritesh, Kumar ;PAUL, Pintu.LIQUID DETERGENT COMPOSITION： WO2014016134A1[P].2014-01-30.